Nonlinear Higher Order Differential and Integral Coupled Systems

Impulsive and Integral Equations on Bounded
and Unbounded Domains

TRENDS IN ABSTRACT AND APPLIED ANALYSIS

ISSN: 2424-8746

Series Editor: John R. Graef
                  The University of Tennessee at Chattanooga, USA

This series will provide state of the art results and applications on current topics in the broad area of Mathematical Analysis. Of a more focused nature than what is usually found in standard textbooks, these volumes will provide researchers and graduate students a path to the research frontiers in an easily accessible manner. In addition to being useful for individual study, they will also be appropriate for use in graduate and advanced undergraduate courses and research seminars. The volumes in this series will not only be of interest to mathematicians but also to scientists in other areas. For more information, please go to http://www.worldscientific.com/series/taaa

Published

Vol. 10   *Nonlinear Higher Order Differential and Integral Coupled Systems:*
          Impulsive and Integral Equations on Bounded and
          Unbounded Domains
          by Feliz Manuel Minhós & Robert de Sousa

Vol. 9   *Boundary Value Problems for Fractional Differential*
          Equations and Systems
          by Bashir Ahmad, Johnny Henderson & Rodica Luca

Vol. 8   *Ordinary Differential Equations and Boundary Value Problems*
          Volume II: Boundary Value Problems
          by John R. Graef, Johnny Henderson, Lingju Kong &
          Xueyan Sherry Liu

Vol. 7   *Ordinary Differential Equations and Boundary Value Problems*
          Volume I: Advanced Ordinary Differential Equations
          by John R. Graef, Johnny Henderson, Lingju Kong &
          Xueyan Sherry Liu

Vol. 6   *The Strong Nonlinear Limit-Point/Limit-Circle Problem*
          by Miroslav Bartušek & John R. Graef

Vol. 5   *Higher Order Boundary Value Problems on Unbounded Domains:*
          Types of Solutions, Functional Problems and Applications
          by Feliz Manuel Minhós & Hugo Carrasco

More information on this series can be found at https://www.worldscientific.com/series/taaa

Trends in Abstract and Applied Analysis
Volume **10**

Nonlinear Higher Order Differential and Integral Coupled Systems

Impulsive and Integral Equations on Bounded and Unbounded Domains

Feliz Manuel Minhós
Universidade de Évora, Portugal

Robert de Sousa
Universidade de Cabo Verde, Cape Verde

World Scientific

NEW JERSEY · LONDON · SINGAPORE · BEIJING · SHANGHAI · HONG KONG · TAIPEI · CHENNAI · TOKYO

Published by

World Scientific Publishing Co. Pte. Ltd.

5 Toh Tuck Link, Singapore 596224

USA office: 27 Warren Street, Suite 401-402, Hackensack, NJ 07601

UK office: 57 Shelton Street, Covent Garden, London WC2H 9HE

Library of Congress Cataloging-in-Publication Data

Names: Minhós, Feliz Manuel, author. | Sousa, Robert de, author.
Title: Nonlinear higher order differential and integral coupled systems : impulsive and
    integral equations on bounded and unbounded domains / Feliz Manuel Minhós,
    Universidade de Évora, Portugal, Robert de Sousa, Universidade de Cabo Verde, Cape Verde.
Description: New Jersey : World Scientific, [2022] | Series: Trends in abstract and applied analysis,
    2424-8746 ; Vol. 10 | Includes bibliographical references and index.
Identifiers: LCCN 2020040733 | ISBN 9789811225123 (hardcover) | ISBN 9789811225130 (ebook) |
    ISBN 9789811225147 (ebook other)
Subjects: LCSH: Boundary value problems. | Integral equations.
Classification: LCC QA379 .M555 2022 | DDC 515/.35--dc23
LC record available at https://lccn.loc.gov/2020040733

British Library Cataloguing-in-Publication Data

A catalogue record for this book is available from the British Library.

For any available supplementary material, please visit
https://www.worldscienti ic.com/worldscibooks/10.1142/11961#t=suppl

Desk Editors: Vishnu Mohan/Kwong Lai Fun

Typeset by Stallion Press
Email: enquiries@stallionpress.com

Printed in Singapore

Preface

Boundary value problems on bounded or unbounded intervals, involving two or more coupled systems of the nonlinear differential equations with full nonlinearities are scary and have a gap in literature.

The present work modestly try to fill this gap.

The systems covered in the work are of two types: systems of differential equations, essentially of the second-order with boundary constraints either in bounded or unbounded intervals presented in several forms and conditions (three points, mixed, with functional dependence, homoclinic and heteroclinic, and systems of integral equations).

The existence, and in some cases, the localization of the solutions are carried out in Banach spaces where adequate norms are considered, following arguments and approaches such as: Schauder's fixed-point theorem or Guo–Krasnosel'skiĭ fixed-point theorem in cones, allied to Green's function or its estimates, lower and upper solutions, convenient truncatures, the Nagumo condition presented in different forms, concept of equiconvergence, Carathéodory functions and sequences.

On the other hand, parallel to the theoretical explanation of this work, there is a range of practical examples and applications involving real phenomena, focusing on the physics, mechanics, biology, forestry, dynamical systems, civil and mechanical engineering.

Contents

List of Figures

Introduction

A system of differential equations is a set of equations, where each one relates some values of the function and its derivatives. When there is an interaction and dependence between different variables on a system of differential equations, it is said that the system is coupled. The phenomena, laws and systems that command the universe are neither independent nor isolated. The interaction or coupling is a fundamental characteristic of everything that surrounds us in the universe.

Natural phenomena are generally nonlinear and are modeled with systems of nonlinear higher order differential equations. In addition, these systems also serve to study and to explain various and important problems in science and engineering, that are not possible to analyze with linear systems.

Like this, the present work focuses on systems of nonlinear higher order differential equations with boundary value problems, where the systems considered are coupled.

The main precursors in the study of the differential equations, according to the history of Mathematics, were Gottfried Leibniz[a], Isaac Newton[b] and the brothers, James Bernoulli[c] and John Bernoulli[d] [84, 144].

---

[a]Gottfried Leibniz (1646–1716), was a German polymath, philosopher, scientist, mathematician, diplomat and librarian.

[b]Isaac Newton (1643–1727) was an English mathematician, astronomer, alchemist, natural philosopher, theologian and scientist, most recognized as a physicist and mathematician.

[c]James Bernoulli (1654–1705), from Switzerland, was the first mathematician to develop infinitesimal calculus beyond what had been done by Newton and Leibniz, applying it to new problems.

[d]John Bernoulli (1667–1748) was a Swiss mathematician. His field of action included variational calculus, physical, physics, chemistry, astronomy, optics, theory of navigation and mathematics.

In 1675, Leibniz was the first to study differential equations when considering and solving the trivial equation

$$\int y\,dy = \frac{1}{2}y^2,$$

thus providing tools such as the signal of integral and inverse problem of tangents. It was Leibniz who discovered the technique of separating variables by studying the solution of the equation $f(x)dx = g(y)dy$, developed the technique to solve the homogeneous equation $dy = f\left(\frac{x}{y}\right)dx$, as well as numerous contributions and applications.

Parallel to Leibniz, Newton had great impact as a contributor and in the history of differential equations. One of Newton's most remarkable and significant contributions in this area was in the study of fluxions and their applications. It was in the study of fluxions that he established: given $f(x, y) = 0$, with x and y functions of t,

$$\frac{\partial f}{\partial x}\frac{dx}{dt} + \frac{\partial f}{\partial y}\frac{dy}{dt} = 0.$$

At the same time, the Bernoulli brothers created several methods (used to date) to solve and boost the development of the differential equations. As an example, we refer the study of the famous brachistochrone.

The determination of solutions for nonlinear higher order systems of differential equation with boundary or initial value problems is very complicated and, in most cases, impossible to determine.

In the literature, discretization methods are usually presented, together with numerical methods for the determination of approximate solutions. Even using numerical methods, there is no guarantee that all equations or systems of nonlinear equations will present stable solutions [46].

Thus, in this work, we present sufficient conditions of existence and, in some cases of localization of solutions for systems of differential equations with several full nonlinearities with boundary values, finite or infinite intervals, with generalized impulses, with multipoint conditions, with Phi-Laplacians, ... Depending on the type of system considered, we will be able to prove the existence of several solutions, especially in bounded and unbounded domains, homoclinics and heteroclinics, and several applications, either to theoretical problems or to real-life phenomena.

The existence, uniqueness or multiplicity of solutions, and their structure depend on the nonlinearity involved and on the type of boundary conditions considered. Thus, the study of the solvability of systems of nonlinear differential equations of higher order considers these two arguments:

nonlinearity and boundary conditions. In the literature, the study has been mostly developed in bounded domains with incomplete nonlinearities, that is, where there is no dependence on all variables and all their derivatives, as can be seen, for example, in $[50, 51, 54, 113\text{--}119, 127, 128, 130, 142]$.

This book intends to contribute to the literature presenting results with complete nonlinearities.

In bounded intervals, a wide variety of methods and techniques can be applied, generally based on compact or completely continuous operators.

In the last decade the theory and the application of the Boundary Value Problems (BVP) have been developed in bounded and unbounded intervals, obtaining sufficient conditions to guarantee, in these cases, the existence and possible multiplicity of bounded or unbounded solutions.

As an example, we refer to $[6]$, where some of the existence conditions in infinite intervals are presented for the problem

$$\begin{cases} x''\phi(t)f(t,x,x') = 0, & 0 < t < \infty, \\ x(0) = 0, & x(t) \text{ bounded on } [0, \infty), \end{cases}$$

with physically reasonable assumptions on ϕ and f, or

$$\begin{cases} \frac{1}{p(t)}\left(p(t)x'\right)' = \phi(t)f(t,x,p(t)x'), \ 0 < t < \infty, \\ -\alpha x(0) + \beta \lim_{t \to 0^+} p(t)x'(t) = c, \ \alpha \geq 0, \ \beta \geq 0, \ c \geq 0 \\ x(t) \text{ bounded on } [0, \infty). \end{cases}$$

We can find many more important contributions in $[6, 28, 39, 82, 104, 111, 142, 160, 161, 172, 185, 214, 216, 248, 252]$.

The previous references contain several and varied examples and applications that illustrate the importance of BVP on unbounded domains, such as the study of symmetrical radial solutions in elliptic nonlinear equations, drain flux theory, physical properties of plasmas, flux study unstable gases through porous media, in the determination of the electric potential in an isolated neutral atom, among others.

However, the theoretical framework for higher order BVP on unbounded domains has not yet been established for a large number of boundary conditions.

Besides, the study of systems of impulsive differential equations in both, bounded and unbounded domains (see $[173]$), is too scarce and has very important accessible applications in phenomena studied in physics, chemistry, population dynamics, biotechnology and economics. For these and other reasons, we have devoted some time to these systems in this work.

The qualitative analysis of the BVP in the real line, namely the existence of homoclinic and heteroclinic solutions, has been restricted to differential equations, where in some cases it is possible to obtain the phase portrait or the graphic of the homoclinic or heteroclinic solution. However, in a system of nonlinear differential equations the graphical or geometric component are lost and the study of the solution is scarce, as far as we know. In this respect the results obtained in this work are completely innovative.

It should also be noted that there are equations that usually arise from the reformulation of boundary value problem associated to partial or ordinary differential equations. These equations are called integral equations. The integral equations are usually given in the implicit form and we study the existence, uniqueness, regularity and stability of solutions. They are essentially relevant in physics and engineering. So, we dedicate the last part of this book to the Hammerstein-type, which is a particular case of integral equations, and some interesting applications.

In order to elaborate all the work, the arguments used are Banach spaces and the corresponding norms, based on Green's function, Guo–Krasnosel'skiĭ fixed-point theorem of compression-expansion cones, Schauder's fixed-point theorem, lower and upper solutions method, truncature technique, Nagumo conditions, L^1-Carathéodory functions and sequences and equiconvergent of the associated operators, L^∞-Carathéodory and kernels functions (see [6, 14, 139, 153, 242, 266]).

The present work is structured into four parts.

Part I – Boundary value problems on bounded domains consists of two chapters:

Chapter 1 – *Third-order three point systems with dependence on the first derivative.* Systems where the nonlinearities can depend on the first derivatives are scarce. This chapter contributes to fill that gap, applying cones theory to the third-order three point boundary value problem. A key point in our method is the fact that the Green's function associated to the linear problem, and its first derivatives, are nonnegative and verify some adequate estimates. The existence of a positive and increasing solution of the third-order three point systems of differential equations with dependence on the first derivative, is obtained by the well-known Guo–Krasnosel'skiĭ theorem on compression-expansion cones. The dependence on the first derivatives is overcome by the construction of an adequate cone and suitable conditions of superlinearity/sublinearity near 0 and $+\infty$. In last section an example illustrates the applicability of the theorem.

Chapter 2 – *Functional coupled systems with full nonlinear terms.* In this chapter we consider the boundary value problem composed by the coupled system constituted by second-order coupled systems with full nonlinearities, together with the functional boundary conditions. More precisely, we explained an existence and localization result and an example to show the applicability of the main theorem. An application to a real phenomenon is shown in the last section: a coupled mass-spring system together with functional behavior at the final instant.

Part II – Coupled systems on unbounded domains is organized as follows:

Chapter 3 – *Second-order coupled systems on the half-line.* Second-order coupled systems on the half-line have many applications in physics, biology, mechanics and among other areas. We use lower and upper solutions method combined with a Nagumo type growth condition, the equiconvergence at infinity to establish an existence and location result for the solutions of the coupled system in the half-line. An application for the existence result is applied to a real phenomena: a predator-prey model. An example is used to show the applicability of the localization part.

Chapter 4 – *Homoclinic solutions for second-order coupled systems.* In this chapter, we apply the fixed point theory, lower and upper solutions method combined with adequate growth assumptions on the nonlinearities, to obtain sufficient conditions for the existence of homoclinic solutions of the coupled system. An application to a real phenomenon is shown in the last section: coupled nonlinear Schrödinger system (NLS) modeling spatial solitons in crystals.

Chapter 5 – *Heteroclinic solutions with phi-Laplacians.* In this chapter, we apply the fixed point theory, to obtain sufficient conditions for the existence of heteroclinic solutions of the coupled system involving phi-Laplacians, assuming some adequate conditions on their inverse and on the asymptotic values $A, B, C, D \in \mathbb{R}$.

Part III – Coupled impulsive systems is divided into three chapters:

Chapter 6 – *Impulsive coupled systems with generalized jump conditions.* We consider a second-order impulsive coupled system with full nonlinearities, mixed boundary conditions and generalized impulsive conditions with dependence on the first derivative. This chapter will establish the existence solution of this problem, illustrate an example and applications to a real phenomena: transverse vibration system of elastically coupled double-string model.

Chapter 7 – *Impulsive coupled systems on the half-line.* Here, we consider the second-order impulsive coupled system in half-line composed by the differential equations. It stands out, Carathéodory functions and sequences, the equiconvergence at each impulsive moment and at infinity to prove an existence result for the impulsive coupled systems with generalized jump conditions in half-line and with full nonlinearities, that depend on the unknown functions and their first derivatives. Finally, an application is presented to a real phenomena: a model of the motion of a spring pendulum.

Chapter 8 – *Localization results for impulsive second-order coupled systems on the half-line.* Two localization results are obtained for coupled systems on the half-line, with fully differential equations, together with generalized impulsive conditions considered on infinite impulsive moments. For the result of more general impulsive conditions, a real phenomena is applied: a logging timber model for removal of trees in cables attached to a helicopter.

Part IV – Coupled systems of integral equations is structured into three chapters:

Chapter 9 – *Coupled systems of Hammerstein integral equations.* Using Guo–Krasnosel'skiĭ compression-expansion theory on cones and positive kernel functions, this chapter deals with generalized coupled systems of integral equations of Hammerstein-type with nonlinearities depending on several derivatives of both variables. Moreover, we underline that both equations and both variables can have different regularity. In the last section an example composed by third and second order nonlinear equations, with three-point boundary conditions, illustrates the applicability of the theorem.

Chapter 10 – *Coupled Hammerstein systems with sign-changing kernels.* The main purpose of this chapter is to overcome the positivity of Green's functions and their derivatives for high-order coupled systems and consequently to extend the range of real applications of these results. So, we consider a generalized coupled systems of two integral equations of Hammerstein-type where the kernel functions may change sign, since they remain positive on some subintervals. The nonlinearities may have discontinuities and, on the other hand, we emphasize that the main existence argument is the Guo–Krasnosel'skiĭ compression-expansion theory in a new type of cone, where some requirements may be satisfied only on some subintervals of the domain. The last section shows an application to a coupled system of beams equations to model the bending of the road-bed and the cable in suspension bridges.

Chapter 11 – *Generalized Hammerstein coupled systems on the real line.* In this chapter, we consider a generalized coupled systems of integral equations of Hammerstein-type on the real line with, eventually, discontinuous nonlinearities. The main existence tool is Schauder's fixed point theorem in the space of bounded and continuous functions with bounded and continuous derivatives on $\mathbb{R}$, combined with the equiconvergence at $\pm\infty$, to recover the compactness of the correspondent operators. Lastly, an application to a real phenomenon is shown: a model to study the deflection of a coupled system of infinite beams.

PART I

Boundary value problems on bounded domains

Introduction

This work will consider the Boundary Value Problem (BVP) composed by a system of differential equations subject to a set of admissibility restrictions that must be respected by the solution set of the system.

Systems of nonlinear equations with boundary conditions in finite intervals are more common in the literature and are generally easier to address in determining or guaranteeing the existence of solutions.

The first boundary value problem at the bounded domain is associated with Dirichlet problem which is formulated as follows: given a closed region Ω with the boundary $\partial\Omega$ and the boundary condition

$$\phi = f(x), \ \ x \in \partial\Omega$$

where $f(x)$ is a continuous function, we are asked to find an harmonic function $\phi(x)$ (for a historical remarks see [64]).

In this part we study problems composed of nonlinear second-order and third-order systems, with complete nonlinearities, and with boundary conditions on bounded intervals, more precisely we seek to guarantee the existence and location of solutions on bounded domains with multi-point conditions, functional boundary conditions and mixed boundary conditions.

The arguments used in this part are based mainly on [25, 99, 169]. We underline:

- The research will focus on the search for sufficient conditions that are required for the nonlinearities in order to guarantee the existence of a non-negative solution for third-order systems. The arguments are based on cones theory for third-order three point boundary value problem.
- Combination of the lower and upper solutions method with fixed point theory.

This first part consists of two chapters which cover the existence and location of coupled systems on bounded intervals.

- In the *first chapter* we consider existence of solutions for nonlinear third-order coupled systems with boundary conditions on the bounded interval $[a, b]$, $a, b \in \mathbb{R}$, $a < b$.
- In the *second chapter*, we study the existence and location of solutions for nonlinear second-order coupled systems on $[a, b]$, $a, b \in \mathbb{R}$, $a < b$, with mixed boundary conditions.

Chapter 1

Third-order three point systems with dependence on the first derivative

The solvability of systems of differential equations of second and higher orders, with different types of boundary conditions has received an increasing interest in recent years. See, for instance, [25, 72, 120, 121, 127, 129, 141, 155, 176] and references therein.

Guezane–Lakoud and Zenkoufi [103], using the Leray–Schauder nonlinear alternative, the Banach contraction theorem and Guo–Krasnosel'skiĭ theorem, study the existence, uniqueness and positivity of solution to the third-order three-point nonhomogeneous boundary value problem

$$u''' + f(t, u(t), u'(t)) = 0, \ t \in (0, 1)$$
$$\alpha u'(1) = \beta u'(\eta), \ u(0) = u'(0) = 0,$$

where $\alpha, \ \beta \in \mathbb{R}_+$, $0 < \eta < 1$ and $f \in C([0, 1) \times [0, \infty) \times [0, \infty), [0, \infty))$

In [169], the authors consider the existence of positive solution

$$\begin{cases} -u''' = a(t)f(t, v) \\ -v''' = b(t)h(t, u) \\ u(0) = u'(0) = 0, \ u'(1) = \alpha u'(\eta) \\ v(0) = v'(0) = 0, v'(1) - \alpha v'(\eta), \end{cases}$$

where $f, \ h \in C([0, 1] \times [0, \infty), [0, \infty))$, $0 < \eta < 1$, $1 < \alpha < \frac{1}{\eta}$, $a(t), b(t) \in C([0, 1], [0, \infty)$ and are not identically zero on $\left[\frac{\eta}{\alpha}, \eta\right]$.

However systems where the nonlinearities can depend on the first derivatives are scarce (see [136]). Motivated by the works referred above, this chapter contributes to fill that gap, applying cones theory to the third-order three point boundary value problem

$$\begin{cases} -u'''(t) = f(t, v(t), v'(t)) \\ -v'''(t) = h(t, u(t), u'(t)) \\ u(0) = u'(0) = 0, \ u'(1) = \alpha u'(\eta) \\ v(0) = v'(0) = 0, \ v'(1) = \alpha v'(\eta). \end{cases} \tag{1.1}$$

The non-negative continuous functions $f, h \in C\left([0, 1] \times [0, +\infty)^2,\right.$ $\left.[0, +\infty)\right)$ verify adequate superlinear and sublinear conditions, $0 < \eta < 1$ and the parameter α is such that $1 < \alpha < \frac{1}{\eta}$. Moreover, this chapter is based in [200].

Third-order differential equations can model various phenomena in physics, biology or physiology such as the flow of a thin film of viscous fluid over a solid surface (see [43, 251]), the solitary waves solution of the Korteweg–de Vries equation (see [174]), or the thyroid-pituitary interaction (see [74]).

A key point in our method is the fact that the Green's function associated to the linear problem and its first derivative are nonnegative and verify some adequate estimates. The existence of a positive and increasing solution of the system (1.1), is obtained by the well-known Guo–Krasnosel'skiĭ theorem on compression-expansion cones. The dependence on the first derivatives is overcome by the construction of an adequate cone and suitable conditions of superlinearity/sublinearity near 0 and $+\infty$.

1.1　Preliminary results

It is clear that the pair of functions $(u(t), v(t)) \in \left(C^3[0, 1], (0, +\infty)\right)^2$ is a solution of problem (1.1) if and only if $(u(t), v(t))$ verify the following system of integral equations

$$\begin{cases} u(t) = \int_0^1 G(t, s) f(s, v(s), v'(s)) ds \\ \\ v(t) = \int_0^1 G(t, s) h(s, u(s), u'(s)) ds, \end{cases} \tag{1.2}$$

where $G(t, s)$ is the Green's function associated to problem (1.1), defined by

$$G(t, s) = \frac{1}{2(1 - \alpha\eta)} \begin{cases} (2ts - s^2)(1 - \alpha\eta) + t^2 s(\alpha - 1), & s \leq \min\{\eta, t\}, \\ t^2(1 - \alpha\eta) + t^2 s(\alpha - 1), & t \leq s \leq \eta, \\ (2ts - s^2)(1 - \alpha\eta) + t^2(\alpha\eta - s), & \eta \leq s \leq t, \\ t^2(1 - s), & \max\{\eta, t\} \leq s. \end{cases} \tag{1.3}$$

The following lemmas provide some properties of the Green's functions and its derivative.

Lemma 1.1 ([170]). *Let* $0 < \eta < 1$ *and* $1 < \alpha < \frac{1}{\eta}$. *Then for any* $(t, s) \in [0, 1] \times [0, 1]$, *we have* $0 \leq G(t, s) \leq g_0(s)$, *where*

$$g_0(s) = \frac{1 + \alpha}{1 - \alpha\eta} s(1 - s).$$

Lemma 1.2 ([170]). *Let* $0 < \eta < 1$ *and* $1 < \alpha < \frac{1}{\eta}$. *Then for any* $(t, s) \in [\frac{\eta}{\alpha}, \eta] \times [0, 1]$, *the Green function* $G(t, s)$ *verifies* $G(t, s) \geq k_0 g_0(s)$, *where*

$$0 < k_0 := \frac{\eta^2}{2\alpha^2(1+\alpha)} \min\{\alpha - 1, 1\} < 1. \tag{1.4}$$

The derivative of G is given by

$$\frac{\partial G}{\partial t}(t, s) = \frac{1}{(1 - \alpha\eta)} \begin{cases} s(1 - \alpha\eta) + ts(\alpha - 1) & s \leq \min\{\eta, t\}, \\ t(1 - \alpha\eta) + ts(\alpha - 1) & t \leq s \leq \eta, \\ s(1 - \alpha\eta) + t(\alpha\eta - s) & \eta \leq s \leq t, \\ t(1 - s) & \max\{\eta, t\} \leq s, \end{cases}$$

and verifies the following lemmas:

Lemma 1.3. *For* $0 < \eta < 1$, $1 < \alpha < \frac{1}{\eta}$ *and any* $(t, s) \in [0, 1] \times [0, 1]$, *we have* $0 \leq \frac{\partial G}{\partial t}(t, s) \leq g_1(s)$, *where*

$$g_1(s) = \frac{(1 - s)}{(1 - \alpha\eta)}.$$

Proof. For $s \leq \min\{\eta, t\}$, we have

$$\frac{t(1 - \alpha\eta) + ts(\alpha - 1)}{(1 - \alpha\eta)} \leq \frac{s(1 - \alpha\eta) + s(\alpha - 1)}{(1 - \alpha\eta)} = \frac{s(\alpha - \alpha\eta)}{(1 - \alpha\eta)}$$

$$= \frac{s\alpha(1 - \eta)}{(1 - \alpha\eta)} \leq \frac{s\alpha(1 - s)}{(1 - \alpha\eta)} \leq \frac{(1 - s)}{(1 - \alpha\eta)}.$$

If $t \leq s \leq \eta$, then

$$\frac{t(1 - \alpha\eta) + ts(\alpha - 1)}{(1 - \alpha\eta)} = \frac{t(1 - \alpha\eta + s\alpha - s)}{(1 - \alpha\eta)} \leq \frac{(1 - \alpha\eta + \eta\alpha - s)}{(1 - \alpha\eta)}$$

$$= \frac{(1 - s)}{(1 - \alpha\eta)}.$$

For $\eta \leq s \leq t$, we have

$$\frac{s(1 - \alpha\eta) + t(\alpha\eta - s)}{(1 - \alpha\eta)} \leq \frac{s(1 - \alpha\eta) + (\alpha\eta - s)}{(1 - \alpha\eta)} = \frac{\alpha\eta(1 - s)}{(1 - \alpha\eta)}$$

$$\leq \frac{\alpha s(1 - s)}{(1 - \alpha\eta)} \leq \frac{(1 - s)}{(1 - \alpha\eta)}.$$

If $\max\{\eta, t\} \leq s$, then

$$\frac{t(1 - s)}{(1 - \alpha\eta)} \leq \frac{s(1 - s)}{(1 - \alpha\eta)} \leq \frac{(1 - s)}{(1 - \alpha\eta)}.$$

So,

$$\frac{\partial G}{\partial t}(t, s) \leq g_1(s) := \frac{(1 - s)}{(1 - \alpha\eta)}, \text{ for } (t, s) \in [0, 1] \times [0, 1]. \qquad \square$$

Lemma 1.4. *For $0 < \eta < 1$, $1 < \alpha < \frac{1}{\eta}$ and any $(t, s) \in [\frac{\eta}{\alpha}, \eta] \times [\frac{\eta}{\alpha}, \eta]$, the derivative of the Green function $\frac{\partial G}{\partial t}(t, s)$ verifies $\frac{\partial G}{\partial t}(t, s) \geq k_1 g_1(s)$, with*

$$0 < k_1 := \min \left\{ \begin{array}{c} \left[\frac{\eta}{\alpha^2}(1 - \alpha\eta) + \frac{\eta^2}{\alpha^3}(\alpha - 1) \right](\alpha - \eta), \\ \frac{\eta(\alpha - \eta)(1 - \alpha\eta)}{\alpha^2} \end{array} \right\}. \tag{1.5}$$

Proof. To find k_1 such that

$$k_1 g_1(s) \leq \frac{\partial G}{\partial t}(t, s),$$

we evaluate it in each branch of $\frac{\partial G}{\partial t}(t, s)$ for $(t, s) \in [\frac{\eta}{\alpha}, \eta] \times [\frac{\eta}{\alpha}, \eta]$,

(i) For $s \leq \min\{\eta, t\}$, we must prove that

$$k_1 \frac{1 - s}{1 - \alpha\eta} \leq \frac{s(1 - \alpha\eta) + ts(\alpha - 1)}{1 - \alpha\eta},$$

that is

$$k_1 \leq \frac{s(1 - \alpha\eta) + ts(\alpha - 1)}{1 - s}.$$

In fact,

$$\frac{s(1 - \alpha\eta) + ts(\alpha - 1)}{1 - s} \geq \frac{\frac{\eta}{\alpha}(1 - \alpha\eta) + \frac{\eta^2}{\alpha^2}(\alpha - 1)}{1 - s}$$

$$\geq \frac{\frac{\eta}{\alpha}(1 - \alpha\eta) + \frac{\eta^2}{\alpha^2}(\alpha - 1)}{\frac{\alpha}{\alpha - \eta}}$$

$$\geq \left[\frac{\eta}{\alpha^2}(1 - \alpha\eta) + \frac{\eta^2}{\alpha^3}(\alpha - 1) \right](\alpha - \eta) \geq k_1 > 0.$$

(ii) If $t \leq s \leq \eta$, the inequality

$$k_1 \frac{1 - s}{1 - \alpha\eta} \leq \frac{t(1 - \alpha\eta + s\alpha - s)}{1 - \alpha\eta}$$

holds for

$$k_1 \leq \frac{t(1 - \alpha\eta + s\alpha - s)}{1 - s}.$$

Therefore, we can take

$$\frac{t(1 - \alpha\eta + s\alpha - s)}{1 - s} \geq \frac{\frac{\eta}{\alpha}\left(1 - \alpha\eta + \frac{\eta\alpha}{\alpha} - \eta\right)}{\frac{\alpha}{\alpha - \eta}}$$

$$= \frac{\eta(\alpha - \eta)(1 - \alpha\eta)}{\alpha^2} \geq k_1 > 0.$$

So, for

$$0 < k_1 := \min \left\{ \begin{array}{c} \left[\frac{\eta}{\alpha^2}(1 - \alpha\eta) + \frac{\eta^2}{\alpha^3}(\alpha - 1) \right](\alpha - \eta), \\ \frac{\eta(\alpha - \eta)(1 - \alpha\eta)}{\alpha^2} \end{array} \right\},$$

we have

$$\frac{\partial G}{\partial t}(t, s) \geq k_1 g_1(s). \qquad \square$$

The existence tool will be the well known Guo-Krasnoselskii result in the expansive and compressive cones theory:

Lemma 1.5 ([107]). *Let $(E, \| \cdot \|)$ be a Banach space, and $P \subset E$ be a cone in E. Assume that Ω_1 and Ω_2 are open subsets of E such that $0 \in \Omega_1$, $\overline{\Omega_1} \subset \Omega_2$.*
  If $T : P \cap (\overline{\Omega_2} \setminus \Omega_1) \to P$ is a completely continuous operator such that either

(i) $\|Tu\| \leq \|u\|$, $u \in P \cap \partial\Omega_1$, *and* $\|Tu\| \geq \|u\|$, $u \in P \cap \partial\Omega_2$,
   or

(ii) $\|Tu\| \geq \|u\|$, $u \in P \cap \partial\Omega_1$, *and* $\|Tu\| \leq \|u\|$, $u \in P \cap \partial\Omega_2$,
   then T has a fixed point in $P \cap (\overline{\Omega_2} \setminus \Omega_1)$.

1.2   Main result

Consider the following growth assumptions:

$(A1)$ $\displaystyle \limsup_{t \in [0,1], \, \|v\|_{C^1} \to 0} \frac{f(t, v, v')}{|v| + |v'|} = 0$ and $\displaystyle \limsup_{t \in [0,1], \, \|u\|_{C^1} \to 0} \frac{h(t, u, u')}{|u| + |u'|} = 0;$

$(A2)$ $\displaystyle \liminf_{t \in [0,1], \, \|v\|_{C^1} \to +\infty} \frac{f(t, v, v')}{|v| + |v'|} = +\infty$ and $\displaystyle \liminf_{t \in [0,1], \, \|u\|_{C^1} \to +\infty} \frac{h(t, u, u')}{|u| + |u'|} = +\infty;$

$(A3)$ $\displaystyle \liminf_{t \in [0,1], \, \|v\|_{C^1} \to 0} \frac{f(t, v, v')}{|v| + |v'|} = +\infty$ and $\displaystyle \liminf_{t \in [0,1], \, \|u\|_{C^1} \to 0} \frac{h(t, u, u')}{|u| + |u'|} = +\infty;$

$(A4)$ $\displaystyle \limsup_{t \in [0,1], \, \|v\|_{C^1} \to +\infty} \frac{f(t, v, v')}{|v| + |v'|} = 0$ and $\displaystyle \limsup_{t \in [0,1], \, \|u\|_{C^1} \to +\infty} \frac{h(t, u, u')}{|u| + |u'|} = 0.$

The main result is given by the following theorem.

Theorem 1.1. *Let $f, h : [0, 1] \times [0, +\infty)^2 \to [0, +\infty)$ be continuous functions such that assumptions $(A1)$ and $(A2)$, or $(A3)$ and $(A4)$, hold. Then* *problem (1.1) has at least one positive solution* $(u(t), v(t)) \in \left(C^3[0, 1]\right)^2$, *that is $u(t) > 0$, $v(t) > 0$, $\forall t \in [0, 1]$.*

Proof. Let $E = C^1[0, 1]$ be the Banach space equipped with the norm $\| \cdot \|_{C^1}$, defined by $\|w\|_{C^1} := \max\left\{\|w\|, \|w'\|\right\}$ and $\|y\| := \max_{t \in [0, 1]} |y(t)|$.

Consider the set

$$K = \left\{ w \in E \ : \ w(t) \geq 0, \quad \min_{t \in \left[\frac{\eta}{\alpha}, \eta\right]} w(t) \geq k_0 \|w\|, \quad \min_{t \in \left[\frac{\eta}{\alpha}, \eta\right]} w'(t) \geq k_1 \|w'\| \right\},$$

with k_0 and k_1 given by (1.4) and (1.5), respectively, and the operators $T_1 : K \to K$ and $T_2 : K \to K$ such that

$$\begin{cases} T_1 u(t) = \int_0^1 G(t, s) f(s, v(s), v'(s)) ds \\[2mm] T_2 v(t) = \int_0^1 G(t, s) h(s, u(s), u'(s)) ds. \end{cases} \tag{1.6}$$

By (1.2), the solutions of the initial system (1.1) are fixed points of the operator $T := (T_1, T_2)$.

First we show that K is a cone. By definition of K it is clear that K is not identically zero or empty.

Consider $a, b \in \mathbb{R}^+$ and $x, y \in K$. Then

$$x \in K \Rightarrow x \in E \ : \ x(t) \geq 0, \quad \min_{t \in [0, 1]} x(t) \geq k_0 \|x\|, \quad \min_{t \in [0, 1]} x'(t) \geq k_1 \|x'\|,$$

$$y \in K \Rightarrow y \in E \ : \ y(t) \geq 0, \quad \min_{t \in [0, 1]} y(t) \geq k \|y\|, \quad \min_{t \in [0, 1]} y'(t) \geq k_1 \|y'\|.$$

As E is a vector space, consider the linear combination $ax + by \in E$.

$$\min_{t \in [0, 1]} (ax(t) + by(t)) = a \min_{t \in [0, 1]} x(t) + b \min_{t \in [0, 1]} y(t)$$
$$\geq ak_0 \|x\| + bk_0 \|y\| = k_0 \left(a\|x\| + b\|y\|\right)$$
$$\geq k_0 \|ax(t) + by(t)\|,$$

and

$$\min_{t \in [0, 1]} (ax(t) + by(t))' = a \min_{t \in [0, 1]} (x(t))' + b \min_{t \in [0, 1]} (y(t))'$$
$$\geq ak_1 \|x'\| + bk_1 \|y'\| = k_1 \left(a\|x'\| + b\|y'\|\right)$$
$$\geq k_0 \|(ax(t) + by(t))'\|.$$

Therefore $ax + by \in K$, that is K is a cone.

Now we show that T_1 and T_2 are completely continuous, i.e., T_1 and T_2 are equicontinuous and uniformly bounded.

For the reader's convenience the proof for T_1 will follow several steps and claims. The arguments for T_2 are analogous.

Step 1: T_1 *and* T_2 *are well defined in* K.

To prove that $T_1 K \subset K$ consider $u \in K$.
As $G(t, s) \geq 0$ for $(t, s) \in [0, 1] \times [0, 1]$, it is clear that $T_1 u(t) \geq 0$.
By Lemma 1.1, the positivity of f and (1.6),

$$0 \leq T_1 u(t) = \int_0^1 G(t, s) f(s, v(s), v'(s)) ds \leq \int_0^1 g_0(s) f(s, v(s), v'(s)) ds.$$

So,

$$\|T_1 u\| \leq \int_0^1 g_0(s) f(s, v(s), v'(s)) ds. \tag{1.7}$$

From Lemma 1.2 and (1.7),

$$T_1 u(t) = \int_0^1 G(t, s) f(s, v(s), v'(s)) ds$$

$$\geq k_0 \int_0^1 g_0(s) f(s, v(s), v'(s)) ds \geq k_0 \|T_1 u\|, \text{ for } t \in \left[\frac{\eta}{\alpha}, \eta\right],$$

with k_0 given by (1.4). By Lemma 1.3,

$$(T_1 u(t))' = \int_0^1 \frac{\partial G}{\partial t}(t, s) f(s, v(s), v'(s)) ds \leq \int_0^1 g_1(s) f(s, v(s), v'(s)) ds.$$

So,

$$\|(T_1 u)'\| \leq \int_0^1 g_1(s) f(s, v(s), v'(s)) ds. \tag{1.8}$$

By Lemma 1.4 and (1.8), it follows

$$(T_1 u(t))' = \int_0^1 \frac{\partial G}{\partial t}(t, s) f(s, v(s), v'(s)) ds$$

$$\geq k_1 \int_0^1 g_1(s) f(s, v(s), v'(s) ds$$

$$\geq k_1 \|(T_1 u)'\|, \text{ for } t \in \left[\frac{\eta}{\alpha}, \eta\right],$$

and k_1 as in (1.5).
So $T_1 K \subset K$. Analogously it can be shown that $T_2 K \subset K$.

Assume that $(A1)$ and $(A2)$ hold.

By $(A1)$, there exists $0 < \delta_1 < 1$ such that, for $(t, v, v') \in [0, 1] \times [0, \delta_1]^2$ and $(t, u, u') \in [0, 1] \times [0, \delta_1]^2$,

$$f(s,\, v(s),\, v'(s)) \leq \varepsilon_1 \left(|v(s)| + |v'(s)|\right) \tag{1.9}$$

and

$$h(s,\, u(s),\, u'(s)) \leq \varepsilon_2 \left(|u(s)| + |u'(s)|\right), \tag{1.10}$$

with ε_1 and ε_2 to be defined forward.

Step 2: *T_1 and T_2 are completely continuous in $C^1[0, 1]$.*

T_1 is continuous in $C^1[0, 1]$ as $G(t, s)$, $\frac{\partial G}{\partial t}(t, s)$ and f are continuous.

Consider the set $B \subset K$, bounded in C^1, and let $u, v \in B$. Then there are $M_1, M_2 > 0$ such that $\|u\|_{C^1} < M_1$ and $\|v\|_{C^1} < M_2$.

Claim 2.1. *T_1 is uniformly bounded in $C^1[0, 1]$.*

In fact, taking $\delta_1 := \min\{M_1, M_2\}$ in (10.10), there are $M_3, M_4 > 0$ such that

$$
\begin{aligned}
\|T_1 u\| &= \max_{t \in [0,\,1]} |T_1 u(t)| = \max_{t \in [0,\,1]} \left| \int_0^1 G(t,s) f(s,\, v(s),\, v'(s)) ds \right| \\
&\leq \int_0^1 \max_{t \in [0,\,1]} |G(t,s)| \, |f(s,\, v(s),\, v'(s))| \, ds \\
&\leq \int_0^1 \max_{t \in [0,\,1]} |G(t,s)| \, \varepsilon_1 \left(|v(s)| + |v'(s)|\right) ds \\
&\leq 2\varepsilon_1 \|v\|_{C^1} \int_0^1 \max_{t \in [0,\,1]} |G(t,s)| < M_3, \ \forall u \in B,
\end{aligned}
$$

$$
\begin{aligned}
\|(T_1 u)'\| &= \max_{t \in [0,\,1]} \left| (T_1 u(t))' \right| = \max_{t \in [0,\,1]} \left| \int_0^1 \frac{\partial G}{\partial t}(t,s) f(s,\, v(s),\, v'(s)) ds \right| \\
&\leq \int_0^1 \max_{t \in [0,\,1]} \left| \frac{\partial G}{\partial t}(t,s) \right| |f(s,\, v(s),\, v'(s))| \, ds \\
&\leq \int_0^1 \max_{t \in [0,\,1]} \left| \frac{\partial G}{\partial t}(t,s) \right| \varepsilon_1 \left(|v(s)| + |v'(s)|\right) ds \\
&\leq 2\varepsilon_1 \|v\|_{C^1} \int_0^1 \max_{t \in [0,\,1]} \left| \frac{\partial G}{\partial t}(t,s) \right| ds < M_4, \ \forall u \in B.
\end{aligned}
$$

Defining $M := \max\{M_3, M_4\}$, then $\|T_1 u\|_{C^1} \leq M$.

Claim 2.2. T_1 *is equicontinuous in* $C^1[0, 1]$.

Let t_1 and $t_2 \in [0, 1]$. Without loss of generality suppose $t_1 \leq t_2$. So

$$
\begin{aligned}
|Tu(t_1) - Tu(t_2)| &= \left| \int_0^1 [G(t_1, s) - G(t_2, s)] f(s, v(s), v'(s)) ds \right| \\
&\leq \int_0^1 |G(t_1, s) - G(t_2, s)| \, \varepsilon_1 \left(|v| + |v'| \right) ds \\
&\leq 2\varepsilon_1 \|v\|_{C^1} \int_0^1 |G(t_1, s) - G(t_2, s)| \, ds \to 0,
\end{aligned}
$$

as $t_1 \to t_2$ and

$$
\begin{aligned}
|(Tu(t_1))' - (Tu(t_2))'| &= \left| \int_0^1 \left[\frac{\partial G}{\partial t}(t_1, s) - \frac{\partial G}{\partial t}(t_2, s) \right] f(s, v(s), v'(s)) ds \right| \\
&\leq \int_0^1 \left| \frac{\partial G}{\partial t}(t_1, s) - \frac{\partial G}{\partial t}(t_2, s) \right| \varepsilon_1 \left(|v| + |v'| \right) ds \\
&\leq 2\varepsilon_1 \|v\|_{C^1} \int_0^1 \left| \frac{\partial G}{\partial t}(t_1, s) - \frac{\partial G}{\partial t}(t_2, s) \right| ds \to 0,
\end{aligned}
$$

as $t_1 \to t_2$.

By the Arzèla-Ascoli Theorem, $T_1 B$ is relatively compact, that is, T_1 is compact.

Applying the same technique, using (1.10), it can be shown that T_2 is compact, too. Consequently T is compact.

Next steps will prove that assumptions of Lemma 1.5 hold.

Step 3: $\|T_1 u\|_{C^1} \leq \|u\|_{C^1}$, *for some* $\rho_1 > 0$ *and* $u \in K \cap \partial\Omega_1$ *with* $\Omega_1 = \{u \in E : \|u\|_{C^1} < \rho_1\}$.

By (A1), define $0 < \rho_1 < 1$ such that $(t, v, v') \in [0, 1] \times [0, \rho_1]^2$ and $(t, u, u') \in [0, 1] \times [0, \rho_1]^2$.

From (10.10) and (1.10), choose $\varepsilon_1, \varepsilon_2 > 0$ sufficiently small such that

$$
\max \left\{
\begin{array}{l}
\varepsilon_1 \varepsilon_2 \displaystyle\int_0^1 g_0(s) ds \int_0^1 (g_0(r) + g_1(r)) \, dr, \\
\varepsilon_1 \varepsilon_2 \displaystyle\int_0^1 g_1(s) ds \int_0^1 (g_0(r) + g_1(r)) \, dr
\end{array}
\right\} < \frac{1}{2}. \tag{1.11}
$$

If $u \in K$ and $\|u\|_{C^1} = \rho_1$, then, by Lemma 1.1, (1.2) and (1.11),

$$
\begin{aligned}
T_1 u(t) &\leq \int_0^1 g_0(s)\varepsilon_1 \left(|v| + |v'|\right) ds \\
&\leq \int_0^1 g_0(s)\varepsilon_1 \int_0^1 \left(|G(t,r)| + \left|\frac{\partial G}{\partial t}(t,r)\right|\right) |h(r, u(r), u'(r))| \, drds \\
&\leq \int_0^1 g_0(s)\varepsilon_1 \int_0^1 (g_0(r) + g_1(r)) |h(r, u(r), u'(r))| \, drds \\
&\leq \varepsilon_1\varepsilon_2 \int_0^1 g_0(s)ds \int_0^1 (g_0(r) + g_1(r)) \left(|u(r)| + |u'(r)|\right) dr \\
&\leq 2\varepsilon_1\varepsilon_2 \|u\|_{C^1} \int_0^1 g_0(s)ds \int_0^1 (g_0(r) + g_1(r)) \, dr < \|u\|_{C^1},
\end{aligned}
$$

and

$$
\begin{aligned}
(T_1 u(t))' &= \int_0^1 \frac{\partial G}{\partial t}(t,s) f(s, v(s), v'(s)) ds \leq \int_0^1 g_1(s)\varepsilon_1 \left(|v| + |v'|\right) ds \\
&\leq \int_0^1 g_1(s)\varepsilon_1 \int_0^1 (g_0(r) + g_1(r)) |h(r, u(r), u'(r))| \, drds \\
&\leq 2\varepsilon_1\varepsilon_2 \|u\|_{C^1} \int_0^1 g_1(s)ds \int_0^1 (g_0(r) + g_1(r)) \, dr < \|u\|_{C^1}.
\end{aligned}
$$

Therefore $\|T_1 u\|_{C^1} \leq \|u\|_{C^1}$.

Step 4: $\|T_1 u\|_{C^1} \geq \|u\|_{C^1}$, *for some* $\rho_2 > 0$ *and* $u \in K \cap \partial\Omega_2$ *with* $\Omega_2 = \{u \in E : \|u\|_{C^1} < \rho_2\}$.

By $(A2)$, $\|v\|_{C^1} \to +\infty$ and $\|u\|_{C^1} \to +\infty$. Therefore there are several cases to be considered:

Case 4.1. *Suppose that there exist* $\theta_1, \theta_2 > 0$ *such that* $\|v\| \to +\infty$, $\|v'\| \leq \theta_1$, $\|u\| \to +\infty$ *and* $\|u'\| \leq \theta_2$.

Consider $\rho > 0$ such that for $(t, v, v') \in [0, 1] \times [\rho, +\infty) \times [0, \theta_1]$ and $(t, u, u') \in [0, 1] \times [\rho, +\infty) \times [0, \theta_2]$, we have

$$
f(t, v(t), v'(t)) \geq \xi_1 \left(|v(t)| + |v'(t)|\right) \tag{1.12}
$$

and

$$
h(t, u(t), u'(t)) \geq \xi_2 \left(|u(t)| + |u'(t)|\right), \tag{1.13}
$$

with ξ_1, ξ_2 such that

$$\min \left\{ \begin{array}{l} (k_0)^2 \, \xi_1 \xi_2 \int_{\frac{\eta}{\alpha}}^{\eta} g_0(s) ds \int_{\frac{\eta}{\alpha}}^{\eta} (k_0 g_0(r) + k_1 g_1(r)) \, dr, \\[2ex] \xi_1 \xi_2 k_0 k_1 \int_{\frac{\eta}{\alpha}}^{\eta} g_0(s) ds \int_{\frac{\eta}{\alpha}}^{\eta} (k_0 g_0(r) + k_1 g_1(r)) \, dr, \\[2ex] \xi_1 \xi_2 k_0 k_1 \int_{\frac{\eta}{\alpha}}^{\eta} g_1(s) ds \int_{\frac{\eta}{\alpha}}^{\eta} (k_0 g_0(r) + k_1 g_1(r)) \, dr, \\[2ex] (k_1)^2 \, \xi_1 \xi_2 \int_{\frac{\eta}{\alpha}}^{\eta} g_1(s) \, ds \int_{\frac{\eta}{\alpha}}^{\eta} (k_0 g_0(r) + k_1 g_1(r)) \, dr \\[2ex] k_0 \, (k_0 + k_1) \, \xi_1 \xi_2 \int_{\frac{\eta}{\alpha}}^{\eta} g_0(s) \, ds \int_{\frac{\eta}{\alpha}}^{\eta} (k_0 g_0(r) + k_1 g_1(r)) \, dr, \\[2ex] k_0 \, (k_0 + k_1) \, \xi_1 \xi_2 \int_{\frac{\eta}{\alpha}}^{\eta} g_1(s) \, ds \int_{\frac{\eta}{\alpha}}^{\eta} (k_0 g_0(r) + k_1 g_1(r)) \, dr \end{array} \right\} > 1,$$

(1.14)

with k_0, k_1 as in (1.4) and (1.5).

Let $u, v \in K$ such that $\|u\|_{C^1} = \rho_2$, where $\rho_2 := \max \left\{ 2\rho_1, \frac{\rho}{k_0}, \frac{\rho}{k_1} \right\}$.

Then $\|u\|_{C^1} = \|u\| = \rho_2$ and $u(t) \geq k_0 \|u\|_{C^1} = k_0 \rho_2 \geq \rho$, $t \in [0, 1]$.
Similarly, $\|v\|_{C^1} = \|v\| = \rho_2$ and $v(t) \geq k_1 \|v\|_{C^1} = k_0 \rho_2 \geq \rho$.

By Lemma 1.2, (1.2) and (9.13),

$$T_1 u(t) \geq \int_{\frac{\eta}{\alpha}}^{\eta} G(t, s) f(s, v(s), v'(s)) ds$$

$$\geq k_0 \int_{\frac{\eta}{\alpha}}^{\eta} g_0(s) \, f(s, v(s), v'(s)) ds \geq k_0 \xi_1 \int_{\frac{\eta}{\alpha}}^{\eta} g_0(s) \, (|v(s)| + |v'(s)|) \, ds$$

$$= k_0 \xi_1 \int_{\frac{\eta}{\alpha}}^{\eta} g_0(s) \, ds \int_{\frac{\eta}{\alpha}}^{\eta} \left(|G(t, r)| + \left| \frac{\partial G}{\partial t}(t, r) \right| \right) |h(r, u(r), u'(r))| \, dr$$

$$\geq k_0 \xi_1 \xi_2 \int_{\frac{\eta}{\alpha}}^{\eta} g_0(s) \, ds \int_{\frac{\eta}{\alpha}}^{\eta} (k_0 g_0(r) + k_1 g_1(r)) \, (|u(r)| + |u'(r)|) \, dr$$

$$= k_0 \xi_1 \int_{\frac{\eta}{\alpha}}^{\eta} g_0(s) \, ds \int_{\frac{\eta}{\alpha}}^{\eta} \left(|G(t, r)| + \left| \frac{\partial G}{\partial t}(t, r) \right| \right) |h(r, u(r), u'(r))| \, dr$$

$$\geq k_0 \xi_1 \xi_2 \int_{\frac{\eta}{\alpha}}^{\eta} g_0(s) \, ds \int_{\frac{\eta}{\alpha}}^{\eta} (k_0 g_0(r) + k_1 g_1(r)) \, (|u(r)| + |u'(r)|) \, dr$$

$$\geq k_0 \xi_1 \xi_2 \int_{\frac{\eta}{\alpha}}^{\eta} g_0(s) \, ds \int_{\frac{\eta}{\alpha}}^{\eta} (k_0 g_0(r) + k_1 g_1(r))$$

$$\times \left(\min_{r \in \left[\frac{\eta}{\alpha}, \eta \right]} u(r) + \min_{r \in \left[\frac{\eta}{\alpha}, \eta \right]} u'(r) \right) dr$$

$$\geq k_0 \xi_1 \xi_2 \int_{\frac{\eta}{\alpha}}^{\eta} g_0(s)\, ds \int_{\frac{\eta}{\alpha}}^{\eta} (k_0 g_0(r) + k_1 g_1(r))\ (k_0 \|u\| + k_1 \|u'\|)\, dr$$

$$\geq k_0 \xi_1 \xi_2 \int_{\frac{\eta}{\alpha}}^{\eta} g_0(s)\, ds \int_{\frac{\eta}{\alpha}}^{\eta} (k_0 g_0(r) + k_1 g_1(r))\ k_0 \|u\|_{C^1}\, dr$$

$$= k_0^2 \|u\|_{C^1} \xi_1 \xi_2 \int_{\frac{\eta}{\alpha}}^{\eta} g_0(s)\, ds \int_{\frac{\eta}{\alpha}}^{\eta} (k_0 g_0(r) + k_1 g_1(r))\, dr > \|u\|_{C^1},$$

and, analogously,

$$(T_1 u(t))' \geq \int_{\frac{\eta}{\alpha}}^{\eta} \frac{\partial G}{\partial t}(t,s) f(s, v(s), v'(s)) ds$$

$$\geq k_1 \int_{\frac{\eta}{\alpha}}^{\eta} g_1(s)\, f(s, v(s), v'(s)) ds$$

$$\geq k_1 \xi_1 \int_{\frac{\eta}{\alpha}}^{\eta} g_1(s)\, (|v(s)| + |v'(s)|)\, ds$$

$$\geq k_1 \xi_1 \xi_2 \int_{\frac{\eta}{\alpha}}^{\eta} g_1(s)\, ds \int_{\frac{\eta}{\alpha}}^{\eta} (k_0 g_0(r) + k_1 g_1(r))\ k_0 \|u\|_{C^1}\, dr$$

$$= k_1 k_0 \|u\|_{C^1} \xi_1 \xi_2 \int_{\frac{\eta}{\alpha}}^{\eta} g_1(s)\, ds \int_{\frac{\eta}{\alpha}}^{\eta} (k_0 g_0(r) + k_1 g_1(r))\, dr > \|u\|_{C^1}.$$

Therefore $\|T_1 u\|_{C^1} \geq \|u\|_{C^1}$.

Case 4.2. *Suppose that there exist $\theta_3, \theta_4 > 0$ such that $\|v'\| \to +\infty$, $\|v\| \leq \theta_3$, $\|u'\| \to +\infty$ and $\|u\| \leq \theta_4$.*

Consider $\rho > 0$ such that for $(t, v, v') \in [0,1] \times [0, \theta_3] \times [\rho, +\infty)$ and $(t, u, u') \in [0,1] \times [0, \theta_4] \times [\rho, +\infty)$, conditions (9.12), (1.13) and (9.13) hold.

Let $u, v \in K$ such that $\|u\|_{C^1} = \rho_2$, where $\rho_2 := \max\left\{2\rho_1, \frac{\rho}{k_0}, \frac{\rho}{k_1}\right\}$.

Then $\|u\|_{C^1} = \|u'\| = \rho_2$ and $u'(t) \geq k_1 \|u'\| = k_1 \rho_2 \geq \rho$, $t \in [0,1]$. Similarly, $\|v\|_{C^1} = \|v'\| = \rho_2$ and $v'(t) \geq k_1 \|v'\| = k_1 \rho_2 \geq \rho$.

As in the previous case, by Lemma 1.2, (1.2) and (9.13)

$$T_1 u(t) \geq \int_{\frac{\eta}{\alpha}}^{\eta} G(t,s) f(s, v(s), v'(s)) ds$$

$$\geq k_0 \int_{\frac{\eta}{\alpha}}^{\eta} g_0(s)\, f(s, v(s), v'(s)) ds \geq k_0 \xi_1 \int_{\frac{\eta}{\alpha}}^{\eta} g_0\, (|v(s)| + |v'(s)|)\, ds$$

$$= k_0 \xi_1 \int_{\frac{\eta}{\alpha}}^{\eta} g_0(s)\, ds \int_{\frac{\eta}{\alpha}}^{\eta} \left(|G(t,r)| + \left|\frac{\partial G}{\partial t}(t,r)\right| \right) |h(r, u(r), u'(r))|\, dr$$

$$= k_1 k_0 \|u\|_{C^1} \xi_1 \xi_2 \int_{\frac{\eta}{\alpha}}^{\eta} g_0(s)\, ds \int_{\frac{\eta}{\alpha}}^{\eta} (k_0 g_0(r) + k_1 g_1(r))\, dr > \|u\|_{C^1},$$

and

$$(T_1 u(t))' \geq \int_{\frac{\eta}{\alpha}}^{\eta} \frac{\partial G}{\partial t}(t, s) f(s, v(s), v'(s)) ds$$

$$\geq k_1 \int_{\frac{\eta}{\alpha}}^{\eta} g_1(s) \, f(s, v(s), v'(s)) ds \geq k_1 \xi_1 \int_{\frac{\eta}{\alpha}}^{\eta} g_1(s) \, (|v(s)| + |v'(s)|) \, ds$$

$$\geq k_1 \xi_1 \xi_2 \int_{\frac{\eta}{\alpha}}^{\eta} g_1(s) \, ds \int_{\frac{\eta}{\alpha}}^{\eta} (k_0 g_0(r) + k_1 g_1(r)) \, k_1 \|u\|_{C^1} dr$$

$$= (k_1)^2 \, \|u\|_{C^1} \xi_1 \xi_2 \int_{\frac{\eta}{\alpha}}^{\eta} g_1(s) \, ds \int_{\frac{\eta}{\alpha}}^{\eta} (k_0 g_0(r) + k_1 g_1(r)) \, dr > \|u\|_{C^1}.$$

Case 4.3. *Suppose that* $\|v\| \to +\infty$, $\|v'\| \to +\infty$, $\|u\| \to +\infty$ *and* $\|u'\| \to +\infty$.

Consider $\rho > 0$ such that for $(t, v, v') \in [0, 1] \times [\rho, +\infty)^2$ and $(t, u, u') \in [0, 1] \times [\rho, +\infty)^2$, conditions (9.12), (1.13) and (9.13) hold.

Let $u, v \in K$ such that $\|u\|_{C^1} = \rho_2$, where $\rho_2 := \max \left\{ 2\rho_1, \frac{\rho}{k_0}, \frac{\rho}{k_1} \right\}$.

Then $\|u\|_{C^1} = \|u\| = \|u'\| = \rho_2$ and $u(t) \geq k_0 \|u\| = k_0 \rho_2 \geq \rho$, $u'(t) \geq k_1 \|u\| = k_1 \rho_2 \geq \rho$, $t \in [0, 1]$. Similarly, $\|v\|_{C^1} = \|v\| = \|v'\| = \rho_2$, $v(t) \geq k_0 \|v\| = k_0 \rho_2 \geq \rho$ and $v'(t) \geq k_1 \|v'\| = k_1 \rho_2 \geq \rho$.

As before,

$$T_1 u(t) \geq k_0 \xi_1 \xi_2 \int_{\frac{\eta}{\alpha}}^{\eta} g_0(s) \, ds \int_{\frac{\eta}{\alpha}}^{\eta} (k_0 g_0(r) + k_1 g_1(r))(|u(r)| + |u'(r)|) \, dr$$

$$\geq k_0 \xi_1 \xi_2 \int_{\frac{\eta}{\alpha}}^{\eta} g_0(s) \, ds \int_{\frac{\eta}{\alpha}}^{\eta} (k_0 g_0(r) + k_1 g_1(r))(k_0 + k_1) \|u\|_{C^1} dr$$

$$= k_0 \, (k_0 + k_1) \, \|u\|_{C^1} \xi_1 \xi_2 \int_{\frac{\eta}{\alpha}}^{\eta} g_0(s) \, ds \int_{\frac{\eta}{\alpha}}^{\eta} (k_0 g_0(r) + k_1 g_1(r)) dr > \|u\|_{C^1},$$

and

$$(T_1 u(t))' \geq k_1 \int_{\frac{\eta}{\alpha}}^{\eta} g_1(s) \, f(s, v(s), v'(s)) ds$$

$$\geq k_1 \xi_1 \int_{\frac{\eta}{\alpha}}^{\eta} g_1(s)(|v(s)| + |v'(s)|) \, ds$$

$$= k_1 \xi_1 \int_{\frac{\eta}{\alpha}}^{\eta} g_1(s) \, ds \int_{\frac{\eta}{\alpha}}^{\eta} \left(|G(t, r)| + \left| \frac{\partial G}{\partial t}(t, r) \right| \right) |h(r, u(r), u'(r))| \, dr$$

$$\geq k_1 \xi_1 \xi_2 \int_{\frac{\eta}{\alpha}}^{\eta} g_1(s) \, ds \int_{\frac{\eta}{\alpha}}^{\eta} (k_0 g_0(r) + k_1 g_1(r))(k_0 + k_1) \|u\|_{C^1} dr$$

$$= k_1 \, (k_0 + k_1) \, \|u\|_{C^1} \xi_1 \xi_2 \int_{\frac{\eta}{\alpha}}^{\eta} g_1(s) \, ds \int_{\frac{\eta}{\alpha}}^{\eta} (k_0 g_0(r) + k_1 g_1(r)) \, dr > \|u\|_{C^1}.$$

The other cases follow the same arguments.

Therefore $\|T_1 u\|_{C^1} \geq \|u\|_{C^1}$.

Then, by Lemma 1.5, T_1 has a fixed point in $K \cap (\overline{\Omega_2} \backslash \Omega_1)$.

By the same steps it can be proved that T_2 has a fixed point in $K \cap (\overline{\Omega_2} \backslash \Omega_1)$, too.

Assume that $(A3)$ and $(A4)$ are verified.

Step 5: $\|T_1 u\|_{C^1} \geq \|u\|_{C^1}$, for some $\rho_3 > 0$ and $u \in K \cap \partial \Omega_3$ with $\Omega_3 = \{u \in E : \|u\|_{C^1} < \rho_3\}$.

By $(A3)$, it can be chosen $\rho_3 > 0$ such that $(t, v, v') \in [0, 1] \times [0, \rho_3]^2$, $(t, u, u') \in [0, 1] \times [0, \rho_3]^2$, and there are $\xi_3, \xi_4 > 0$ with

$$f(t, v(t), v(t)) \geq \xi_3 \left(|v(t)| + |v'(t)|| \right),$$
$$h(t, u(t), u'(t)) \geq \xi_4 \left(|u(t)| + |u'(t)|| \right)$$

and

$$\min \left\{ \begin{array}{l} (k_0)^2 \, \xi_3 \xi_4 \int_{\frac{\eta}{\alpha}}^{\eta} g_0(s) ds \int_{\frac{\eta}{\alpha}}^{\eta} (k_0 g_0(r) + k_1 g_1(r)) \, dr, \\[2mm] k_0 k_1 \xi_3 \xi_4 \int_{\frac{\eta}{\alpha}}^{\eta} g_1(s) \, ds \int_{\frac{\eta}{\alpha}}^{\eta} (k_0 g_0(r) + k_1 g_1(r)) \, dr \\[2mm] \xi_3 \xi_4 k_0 k_1 \int_{\frac{\eta}{\alpha}}^{\eta} g_0(s) \, ds \int_{\frac{\eta}{\alpha}}^{\eta} (k_0 g_0(r) + k_1 g_1(r)) \, dr \\[2mm] (k_1)^2 \, \xi_3 \xi_4 \int_{\frac{\eta}{\alpha}}^{\eta} g_1(s) \, ds \int_{\frac{\eta}{\alpha}}^{\eta} (k_0 g_0(r) + k_1 g_1(r)) \, dr \\[2mm] \xi_3 \xi_4 k_0 \, (k_0 + k_1) \int_{\frac{\eta}{\alpha}}^{\eta} g_0(s) \, ds \int_{\frac{\eta}{\alpha}}^{\eta} (k_0 g_0(r) + k_1 g_1(r)) \, dr \\[2mm] k_1 \, (k_0 + k_1) \, \xi_3 \xi_4 \int_{\frac{\eta}{\alpha}}^{\eta} g_1(s) \, ds \int_{\frac{\eta}{\alpha}}^{\eta} (k_0 g_0(r) + k_1 g_1(r)) \, dr \end{array} \right\} > 1.$$

$$\tag{1.15}$$

Let $u \in K$ and $\|u\|_{C^1} = \rho_3$.

Case 5.1. *Suppose* $\|u\|_{C^1} = \|u\| = \rho_3$.

By Lemma 1.2, (1.2) and (1.15),

$$T_1 u(t) \geq \xi_3 \int_{\frac{\eta}{\alpha}}^{\eta} G(t, s) \left(|v(s)| + |v'(s)| \right) ds$$

$$\geq \xi_3 \xi_4 k_0 \int_{\frac{\eta}{\alpha}}^{\eta} g_0(s) \, ds \int_{\frac{\eta}{\alpha}}^{\eta} (k_0 g_0(r) + k_1 g_1(r)) \, k_0 \|u\|_{C^1} dr$$

$$= \xi_3 \xi_4 \, (k_0)^2 \|u\|_{C^1} \int_{\frac{\eta}{\alpha}}^{\eta} g_0(s) \, ds \int_{\frac{\eta}{\alpha}}^{\eta} (k_0 g_0(r) + k_1 g_1(r)) \, dr > \|u\|_{C^1},$$

and

$$(T_1 u(t))' \geq k_1 \int_{\frac{\eta}{\alpha}}^{\eta} g_1(s) \, f(s, \, v(s), \, v'(s)) ds$$

$$\geq k_1 \xi_3 \int_{\frac{\eta}{\alpha}}^{\eta} g_1(s) \, (|v(s)| + |v'(s)|) \, ds$$

$$\geq k_1 \xi_3 \xi_4 \int_{\frac{\eta}{\alpha}}^{\eta} g_1(s) \, ds \int_{\frac{\eta}{\alpha}}^{\eta} (k_0 g_0(r) + k_1 g_1(r)) \, k_0 \|u\|_{C^1} dr$$

$$= k_0 k_1 \|u\|_{C^1} \xi_3 \xi_4 \int_{\frac{\eta}{\alpha}}^{\eta} g_1(s) \, ds \int_{\frac{\eta}{\alpha}}^{\eta} (k_0 g_0(r) + k_1 g_1(r)) \, dr > \|u\|_{C^1}.$$

Case 5.2. *Suppose* $\|u\|_{C^1} = \|u'\| = \rho_3$.

By Lemma 1.2, (1.2) and (1.15)

$$T_1 u(t) \geq \xi_3 \int_{\frac{\eta}{\alpha}}^{\eta} G(t, s) \, (|v(s)| + |v'(s)|) \, ds$$

$$\geq \xi_3 k_0 \int_{\frac{\eta}{\alpha}}^{\eta} g_0(s) \int_{\frac{\eta}{\alpha}}^{\eta} \left(|G(t, r)| + \left| \frac{\partial G}{\partial t}(t, r) \right| \right) |h(r, u(r), u'(r))| \, dr ds$$

$$\geq \xi_3 \xi_4 k_0 \int_{\frac{\eta}{\alpha}}^{\eta} g_0(s) \, ds \int_{\frac{\eta}{\alpha}}^{\eta} (k_0 g_0(r) + k_1 g_1(r)) \, k_1 \|u\|_{C^1} dr$$

$$= \xi_3 \xi_4 k_0 k_1 \|u\|_{C^1} \int_{\frac{\eta}{\alpha}}^{\eta} g_0(s) \, ds \int_{\frac{\eta}{\alpha}}^{\eta} (k_0 g_0(r) + k_1 g_1(r)) \, dr > \|u\|_{C^1},$$

and

$$(T_1 u(t))' \geq k_1 \xi_3 \xi_4 \int_{\frac{\eta}{\alpha}}^{\eta} g_1(s) \, ds \int_{\frac{\eta}{\alpha}}^{\eta} (k_0 g_0(r) + k_1 g_1(r))$$

$$\times (|u(r)| + |u'(r)|) \, dr$$

$$\geq k_1 \xi_3 \xi_4 \int_{\frac{\eta}{\alpha}}^{\eta} g_1(s) \, ds \int_{\frac{\eta}{\alpha}}^{\eta} (k_0 g_0(r) + k_1 g_1(r))$$

$$\times \left(\min_{r \in [\frac{\eta}{\alpha}, \eta]} u(r) + \min_{r \in [\frac{\eta}{\alpha}, \eta]} u'(r) \right) dr$$

$$= (k_1)^2 \|u\|_{C^1} \xi_3 \xi_4 \int_{\frac{\eta}{\alpha}}^{\eta} g_1(s) \, ds \int_{\frac{\eta}{\alpha}}^{\eta} (k_0 g_0(r) + k_1 g_1(r)) \, dr > \|u\|_{C^1}.$$

Case 5.3. *Suppose* $\|u\|_{C^1} = \|u\| = \|u'\| = \rho_3$.

By Lemma 1.2, (1.2) and (1.15)

$$T_1 u(t) \geq \int_{\frac{\eta}{\alpha}}^{\eta} G(t, s) f(s, v(s), v'(s)) ds$$

$$\geq \xi_3 k_0 \int_{\frac{\eta}{\alpha}}^{\eta} g_0(s) \int_{\frac{\eta}{\alpha}}^{\eta} \left(|G(t, r)| + \left| \frac{\partial G}{\partial t}(t, r) \right| \right) |h(r, u(r), u'(r))| \, dr ds$$

$$\geq \xi_3 \xi_4 k_0 \int_{\frac{\eta}{\alpha}}^{\eta} g_0(s) \, ds \int_{\frac{\eta}{\alpha}}^{\eta} (k_0 g_0(r) + k_1 g_1(r)) \; (k_0 \|u\| + k_1 \|u'\|) \, dr$$

$$= \xi_3 \xi_4 k_0 (k_0 + k_1) \|u\|_{C^1} \int_{\frac{\eta}{\alpha}}^{\eta} g_0(s) \, ds \int_{\frac{\eta}{\alpha}}^{\eta} (k_0 g_0(r) + k_1 g_1(r)) \, dr > \|u\|_{C^1},$$

and

$$(T_1 u(t))' \geq k_1 \int_{\frac{\eta}{\alpha}}^{\eta} g_1(s) \; f(s, v(s), v'(s)) ds$$

$$\geq k_1 \xi_3 \int_{\frac{\eta}{\alpha}}^{\eta} g_1(s) \; (|v(s)| + |v'(s)|) \, ds$$

$$\geq k_1 \xi_3 \xi_4 \int_{\frac{\eta}{\alpha}}^{\eta} g_1(s) \, ds \int_{\frac{\eta}{\alpha}}^{\eta} (k_0 g_0(r) + k_1 g_1(r))$$

$$\times \left(\min_{r \in [\frac{\eta}{\alpha}, \eta]} u(r) + \min_{r \in [\frac{\eta}{\alpha}, \eta]} u'(r) \right) dr$$

$$\geq k_1 (k_0 + k_1) \|u\|_{C^1} \xi_3 \xi_4 \int_{\frac{\eta}{\alpha}}^{\eta} g_1(s) \, ds \int_{\frac{\eta}{\alpha}}^{\eta} (k_0 g_0(r) + k_1 g_1(r)) \; dr > \|u\|_{C^1}.$$

In any case, $\|T_1 u\|_{C^1} \geq \|u\|_{C^1}$.

Step 6: $\|T_1 u\|_{C^1} \leq \|u\|_{C^1}$, *for some* $\rho_4 > 0$ *and* $u \in K \cap \partial \Omega_4$ *with* $\Omega_4 = \{u \in E : \|u\|_{C^1} < \rho_4\}$.

Let $u \in K$ and $\|u\|_{C^1} = \rho_4$.

Case 6.1. *Suppose that f and h are bounded.*

Then there is $N > 0$ such that $f(t, v(t), v'(t)) \leq N$, $h(t, u(t), u'(t)) \leq N$, $\forall u, v \in [0, \infty)$.

Choose

$$\rho_4 = \max \left\{ 2 \rho_3, \; N \int_0^1 g_0(s) ds, \; N \int_0^1 g_1(s) ds \right\}.$$

Then

$$T_1 u(t) = \int_0^1 G(t, s) f(s, v(s), v'(s)) ds \leq N \int_0^1 g_0(s) ds \leq \rho_4, \text{ for } t \in [0, 1],$$

and

$$(T_1 u(t))' = \int_0^1 \frac{\partial G}{\partial t}(t, s) f(s, v(s), v'(s)) ds$$

$$\leq N \int_0^1 g_1(s) ds \leq \rho_4, \text{ for } t \in [0, 1].$$

Thus, $\|T_1 u\|_{C^1} \leq \|u\|_{C^1}$. Similarly $\|T_2 v\|_{C^1} \leq \|v\|_{C^1}$ for any $v \in K$ and $\|v\|_{C^1} = \rho_4$.

Case 6.2. *Consider that f is bounded and h is unbounded.*

So, there is $N > 0$ such that $f(t, v(t), v'(t)) \leq N$, $\forall (v, v') \in [0, +\infty)^2$.

By $(A4)$, there exists $M > 0$ such that $h(t, u(t), u'(t)) \leq \mu(|u(t)| + |u'(t)|)$, whenever $|u(t)| + |u'(t)| \geq M$, with μ verifying

$$\max\left\{ \mu \int_0^1 g_0(s) ds, \mu \int_0^1 g_1(s) ds \right\} < \frac{1}{2}. \tag{1.16}$$

Setting

$$p(r) := \max\{h(t, u(t), u'(t)) : t \in [0, 1], 0 \leq u \leq r, 0 \leq u' \leq r\},$$

we have

$$\lim_{r \to \infty} p(r) = +\infty.$$

Define

$$\rho_4 = \max\left\{ 2\rho_3, M, N \int_0^1 g_0(s) ds, N \int_0^1 g_1(s) ds \right\} \tag{1.17}$$

such that $p(\rho_4) \geq p(r)$, $0 \leq r \leq \rho_4$. Then

$$T_1 u(t) = \int_0^1 G(t, s) f(s, v(s), v'(s)) ds \leq N \int_0^1 g_0(s) ds \leq \rho_4,$$

and

$$(T_1 u(t))' = \int_0^1 \frac{\partial G}{\partial t}(t, s) f(s, v(s), v'(s)) ds \leq N \int_0^1 g_1(s) ds \leq \rho_4, \text{ for } t \in [0, 1].$$

So, $\|T_1 u\|_{C^1} \leq \|u\|_{C^1}$ for $\|u\|_{C^1} = \rho_4$.

Moreover, if $v \in K$ such that $\|v\|_{C^1} = \rho_4$, we have $|u(t)| + |u'(t)| \geq \rho_4 \geq M$,

$$h(t, u(t), u'(t)) \leq \mu\left(|u(t)| + |u'(t)|\right) \leq 2\mu\rho_4, \tag{1.18}$$

and $p(\rho_4) \leq 2\mu\rho_4$. Therefore

$$T_2 v(t) = \int_0^1 G(t, s) h(s, u(s), u'(s)) ds \leq \int_0^1 g_0(s) p(\rho_4) ds$$

$$\leq p(\rho_4) \int_0^1 g_0(s) ds \leq 2\mu\rho_4 \int_0^1 g_0(s) ds \cdot \leq \rho_4,$$

and

$$(T_2v(t))' = \int_0^1 \frac{\partial G}{\partial t}(t,s)h(s,u(s),u'(s))ds \leq \int_0^1 g_1(s)p(\rho_4)ds$$

$$\leq p(\rho_4)\int_0^1 g_1(s)ds \leq 2\mu\rho_4\int_0^1 g_1(s)ds \leq \rho_4.$$

Consequently $\|T_2v\|_{C^1} \leq \|v\|_{C^1}$ for $\|v\|_{C^1} = \rho_4$.

Case 6.3. *Suppose that f is unbounded and h is bounded.*

Then, there is $N > 0$ such that $h(t,u(t),u'(t)) \leq N$, $\forall (u,u') \in [0,+\infty)^2$ and, by $(A4)$, there exists $M > 0$ such that $f(t,v(t),v'(t)) \leq \mu(|v| + |v'|)$, for $(|v| + |v'|) \geq M$ with μ satisfying (9.15).

Choosing ρ_4 as in (1.17), the arguments follow like in the previous case.

Case 6.4. *Consider that f and h are unbounded.*

By $(A4)$, there is $M > 0$ such that $f(t,v(t),v'(t)) \leq \mu(|v| + |v'|)$, $h(t,u(t),u'(t)) \leq \mu(|u| + |u'|)$ for $|v| + |v'| \geq M$ and $|u| + |u'| \geq M$ with μ as in (9.15).

Setting

$$p(r) := \max\{h(t,u(t),u'(t)) : t \in [0,1],\ 0 \leq u \leq r,\ 0 \leq u' \leq r\},$$
$$q(r) := \max\{f(t,v(t),v'(t)) : t \in [0,1],\ 0 \leq v \leq r,\ 0 \leq v' \leq r\}$$

we have

$$\lim_{r\to\infty} p(r) = +\infty \text{ and } \lim_{r\to\infty} q(r) = +\infty.$$

Choose

$$\rho_4 = \max\{2\rho_3,\ M\}$$

such that $p(\rho_4) \geq p(r)$ and $q(\rho_4) \geq q(r)$ for $0 \leq r \leq \rho_4$.

Let $u,v \in K$ and $\|u\|_{C^1} = \|v\|_{C^1} = \rho_4$.

Arguing as in (9.17) it can be easily shown that $\|T_1u\|_{C^1} \leq \|u\|_{C^1}$, $\|T_2v\|_{C^1} \leq \|v\|_{C^1}$.

By Lemma 1.5 the operators T_1, T_2 has a fixed point in $K \cap (\overline{\Omega_4}\backslash\Omega_3)$, therefore $T = (T_1, T_2)$ has a fixed point (u,v) which is a positive solution of the initial problem.

Moreover these functions u and v are given by

$$\begin{cases} u(t) = \int_0^1 G(t,s)f(s,v(s),v'(s))ds \\[2mm] v(t) = \int_0^1 G(t,s)h(s,u(s),u'(s))ds, \end{cases}$$

and are both increasing functions. $\qquad\qquad\square$

1.3 Example

Consider the following third-order nonlinear system

$$
\begin{cases}
-u'''(t) = \left(t^2 + 1\right)\left(e^{-v(t)} + \sqrt{|v'(t)|}\right) \\[2mm]
-v'''(t) = (\,u(t) + 1)^2 \arctan\left(|u'(t)| + 1\right) \\[2mm]
u(0) = u'(0) = 0, u'(1) = \frac{3}{2}u'\left(\frac{1}{2}\right) \\[2mm]
v(0) = v'(0) = 0, v'(1)\grave{} = \frac{3}{2}v'\left(\frac{1}{2}\right).
\end{cases}
\tag{1.19}
$$

In fact this problem is a particular case of system (1.1) with

$$
f(t,\, v(t),\, v'(t)) := \left(t^2 + 1\right)\left(e^{-v(t)} + \sqrt{|v'(t)|}\right)
$$

$$
h(t,\, u(t),\, u'(t)) := (\,u(t) + 1)^2 \arctan\left(|u'(t)| + 1\right),
$$

$$
\eta = \frac{1}{2} \text{ and } \alpha = \frac{3}{2}.
$$

It can be easily check that the above functions are non-negative and verify the assumptions $(A3)$ and $(A4)$.

The functions f and h are non-negative, because they are products of non negative functions. Note that, $\forall t \in [0,1]$ and $\forall(u(t),\, v(t)) \in \left(C^3[0,1],\, (0, +\infty)\right)$, $(t^2 + 1) \geq 1$, $e^{-v(t)} = \frac{1}{e^{v(t)}} \geq 0$, $\sqrt{|v'(t)|} \geq 0$, $(\,u(t) + 1)^2 \geq 1$ and for definition of arctan $: \mathbb{R} \to \left]-\frac{\pi}{2},\, \frac{\pi}{2}\right[$ is such that $\arctan(x) \to \frac{\pi}{2}$, as $x \to +\infty$ and $\arctan(x) \to -\frac{\pi}{2}$, as $x \to -\infty$. So, $f \geq 0$, $h \geq 0$.

Finally, as

$$
\liminf_{t\in[0,1],\, \|v\|_{C^1}\to 0} \frac{\left(t^2 + 1\right)\left(e^{-v(t)} + \sqrt{|v'(t)|}\right)}{|v| + |v'|} = +\infty,
$$

$$
\liminf_{t\in[0,1],\, \|u\|_{C^1}\to 0} \frac{(\,u(t) + 1)^2 \arctan\left(|u'(t)| + 1\right)}{|u| + |u'|} = +\infty,
$$

condition $(A3)$ holds and

$$
\limsup_{t\in[0,1],\, \|v\|_{C^1}\to+\infty} \frac{\left(t^2 + 1\right)\left(e^{-v(t)} + \sqrt{|v'(t)|}\right)}{|v| + |v'|} = 0,
$$

$$
\limsup_{t\in[0,1],\, \|u\|_{C^1}\to+\infty} \frac{(\,u(t) + 1)^2 \arctan\left(|u'(t)| + 1\right)}{|u| + |u'|} = 0,
$$

assumption $(A4)$ is satisfied.

Therefore, by Theorem 1.1, problem (1.19) has at least a positive solution $(u(t),\, v(t)) \in \left(C^3[0,1]\right)^2$, that is $u(t) > 0$, $v(t) > 0$, $\forall t \in [0,1]$.

Chapter 2

Functional coupled systems with full nonlinear terms

In this chapter we consider the boundary value problem composed by the coupled system of the second-order differential equations with full nonlinearities

$$\begin{cases} u''(t) = f(t, u(t), v(t), u'(t), v'(t)), \ t \in [a, b], \\ v''(t) = h(t, u(t), v(t), u'(t), v'(t)), \end{cases} \tag{2.1}$$

where $f, h : [a, b] \times \mathbb{R}^4 \to \mathbb{R}$ are continuous functions, and the functional boundary conditions

$$\begin{cases} u(a) = v(a) = 0 \\ L_1(u, u(b), u'(b)) = 0 \\ L_2(v, v(b), v'(b)) = 0, \end{cases} \tag{2.2}$$

where $L_1, L_2 : C[a, b] \times \mathbb{R}^2 \to \mathbb{R}$ are continuous functions verifying some monotone assumptions.

Ordinary differential systems have been studied by many authors, like, for instance, $[18, 25, 120, 121, 127, 136, 141, 169, 181, 200, 202, 262, 274]$. In particular, second-order coupled systems can be applied to several real phenomena, such as, Lotka-Volterra models, reaction diffusion processes, prey-predator or other interaction systems, Sturm-Liouville problems, mathematical biology, chemical systems (see, for example, $[16, 21, 22, 159, 186, 254]$ and the references therein).

In $[247]$ the authors study the existence of solutions for the nonlinear second-order coupled system

$$\begin{cases} -u''(t) = f_1(t, v(t)), \\ -v''(t) = f_2(t, v(t)), \end{cases} \ t \in [0, 1],$$

with $f_1, f_2 : [0, 1] \times \mathbb{R} \to \mathbb{R}$ are continuous functions, together with the nonlinear boundary conditions

$$\varphi(u(0), v(0), u'(0), v'(0), u'(1), v'(1)) = (0, 0),$$

$$\psi(u(0), v(0)) + (u(1), v(1)) = (0, 0),$$

where $\varphi : \mathbb{R}^6 \to \mathbb{R}^2$ and $\psi : \mathbb{R}^2 \to \mathbb{R}^2$ are continuous functions.

In [72] it provided some growth conditions on the nonnegative nonlinearities of the system

$$\begin{cases} -x''(t) = f_1(t, x(t), y(t)) \\ -y''(t) = f_2(t, x(t), y(t)), \quad t \in (0, 1), \\ x(0) = y(0) = 0, \\ x(1) = \alpha\,[y], \\ y(1) = \beta\,[x], \end{cases}$$

where $f_1, f_2 : (0, 1) \times [0, +\infty)^2 \to [0, +\infty)$ are continuous and may be singular at $t = 0, 1$, and $\alpha[x]$, $\beta[x]$ are bounded linear functionals on $C[0, 1]$ given by

$$\alpha\,[y] = \int_0^1 y(t) dA(t), \ \ \beta\,[x] = \int_0^1 x(t) dB(t),$$

involving Stieltjes integrals, and A, B are functions of bounded variation with positive measures.

Motivated by these works we consider the second-order coupled fully differential equations (2.1) together with the functional boundary conditions (2.2). This chapter is based on [205], and to the best of our knowledge, it is the first time where these coupled differential systems embrace functional boundary conditions. Remark that, the functional dependence includes and generalizes the classical boundary conditions such as separated, multi-point, nonlocal, integro-differential, with maximum or minimum arguments,... More details on such conditions and their potentialities can be seen, for instance, in [50, 98, 102, 203, 267] and the references therein. Our main result is applied to a coupled mass-spring system subject to a new type of global boundary data.

The arguments in this chapter follow lower and upper solutions method and fixed point theory. Therefore, the main result is an existence and localization theorem, as it provides not only the existence of solution, but a strip where the solution varies, as well. Due to an adequate auxiliary problem, including a convenient truncature, there is no need of sign, bound, monotonicity or other growth assumptions on the nonlinearities, besides the Nagumo condition.

2.1   Definitions and preliminaries

Let $E = C^1[a, b]$ be the Banach space equipped with the norm $\|\cdot\|_{C^1}$, defined by

$$\|w\|_{C^1} := \max\{\|w\|, \|w'\|\},$$

where

$$\|y\| := \max_{t \in [a,b]} |y(t)|$$

and $E^2 = \left(C^1\,[a,b]\right)^2$ with the norm

$$\|(u,v)\|_{E^2} = \|u\|_{C^1} + \|v\|_{C^1}.$$

Forward in this work, we consider the following assumption

(A) The functions $L_1, L_2 : C\,[a,b] \times \mathbb{R}^2 \to \mathbb{R}$ are continuous, nonincreasing in the first variable and nondecreasing in the second one.

To apply lower and upper solutions method we consider the next definition:

Definition 2.1. A pair of functions $(\alpha_1, \alpha_2) \in \left(C^2\,[a,b]\right)^2$ is a coupled lower solution of problem (2.1), (2.2) if

$$\alpha_1''(t) \leq f(t, \alpha_1(t), \alpha_2(t), \alpha_1'(t), \alpha_2'(t))$$
$$\alpha_2''(t) \leq h(t, \alpha_1(t), \alpha_2(t), \alpha_1'(t), \alpha_2'(t))$$
$$\alpha_1(a) \leq 0$$
$$\alpha_2(a) \leq 0$$
$$L_1(\alpha_1, \alpha_1(b), \alpha_1'(b)) \geq 0$$
$$L_2(\alpha_2, \alpha_2(b), \alpha_2'(b)) \geq 0.$$

A pair of functions $(\beta_1, \beta_2) \in \left(C^2\,[a,b]\right)^2$ is a coupled upper solution of problem (2.1), (2.2) if it verifies the reverse inequalities.

The Nagumo-type conditions are useful to obtain *a priori* bounds on the first derivatives of the unknown functions:

Definition 2.2. Let $\alpha_1(t), \beta_1(t), \alpha_2(t)$ and $\beta_2(t)$ be continuous functions such that

$$\alpha_1(t) \leq \beta_1(t), \quad \alpha_2(t) \leq \beta_2(t), \quad \forall t \in [a,b].$$

The continuous functions $f, h : [a,b] \times \mathbb{R}^4 \to \mathbb{R}$ satisfy Nagumo-type conditions relative to intervals $[\alpha_1(t), \beta_1(t)]$ and $[\alpha_2(t), \beta_2(t)]$, if, there are $N_1 > r_1$, $N_2 > r_2$, with

$$r_1 := \max\left\{ \frac{\beta_1(b) - \alpha_1(a)}{b-a}, \; \frac{\beta_1(a) - \alpha_1(b)}{b-a} \right\}, \tag{2.3}$$

$$r_2 := \max\left\{ \frac{\beta_2(b) - \alpha_2(a)}{b-a}, \; \frac{\beta_2(a) - \alpha_2(b)}{b-a} \right\}, \tag{2.4}$$

and continuous positive functions $\varphi, \psi : [0, +\infty) \to (0, +\infty)$, such that

$$|f(t, x, y, z, w)| \leq \varphi(|z|), \tag{2.5}$$

for

$$\alpha_1(t) \leq x \leq \beta_1(t), \alpha_2(t) \leq y \leq \beta_2(t), \ \forall t \in [a, b], \text{and } w \in \mathbb{R}, \tag{2.6}$$

and

$$|h(t, x, y, z, w)| \leq \psi(|w|), \tag{2.7}$$

for

$$\alpha_1(t) \leq x \leq \beta_1(t), \alpha_2(t) \leq y \leq \beta_2(t), \ \forall t \in [a, b], \text{and } z \in \mathbb{R}, \tag{2.8}$$

verifying

$$\int_{r_1}^{N_1} \frac{ds}{\varphi(s)} > b - a, \int_{r_2}^{N_2} \frac{ds}{\psi(s)} > b - a. \tag{2.9}$$

Lemma 2.1. *Let $f, h : [a, b] \times \mathbb{R}^4 \to \mathbb{R}$ be continuous functions satisfying Nagumo-type conditions to (2.5) in (2.6) and (2.7) in (2.8).*
Then for every solution $(u, v) \in \left(C^2[a, b]\right)^2$ verifying (2.6) and (2.8), there are $N_1, N_2 > 0$, given by (2.9), such that

$$\|u'\| \leq N_1 \text{ and } \|v'\| \leq N_2. \tag{2.10}$$

Proof. Let $(u(t), v(t))$ be a solution of (2.1) satisfying (2.6).

By Lagrange Theorem, there are $t_0, t_1 \in [a, b]$ such that

$$u'(t_0) = \frac{u(b) - u(a)}{b - a} \text{ and } v'(t_1) = \frac{v(b) - v(a)}{b - a}.$$

Suppose, by contradiction, that $|u'(t)| > r_1, \forall t \in [a, b]$, with r_1 given by (2.3). If $u'(t) > r_1, \forall t \in [a, b]$, by (2.6) we obtain the following contradiction with (2.3):

$$u'(t_0) = \frac{u(b) - u(a)}{b - a} \leq \frac{\beta_1(b) - \alpha_1(a)}{b - a} \leq r_1.$$

If $u'(t) < -r_1, \forall t \in [a, b]$, the contradiction is similar.
In the case where $|u'(t)| \leq r_1, \forall t \in [a, b]$, the proof will be finished.
So, assume that there are $t_2, t_3 \in [a, b]$ such that $t_2 < t_3$, and

$$u'(t_2) < r_1 \text{ and } u'(t_3) > r_1.$$

By continuity, there is $t_4 \in [t_2, t_3]$ such that

$$u'(t_4) = r_1 \text{ and } u'(t_3) > r_1, \ \forall t \in]t_4, t_3].$$

So, by a convenient change of variable, by (2.5), (2.6), and (2.9), we obtain

$$
\int_{u'(t_4)}^{u'(t_3)} \frac{ds}{\varphi(|s|)} = \int_{t_4}^{t_3} \frac{u''(t)}{\varphi(|u'(t)|)} dt \le \int_a^b \frac{|u''(t)|}{\varphi(|u'(t)|)} dt
$$
$$
= \int_a^b \frac{|f(t, u(t), v(t), u'(t), v'(t)|}{\varphi(|u'(t)|)} dt \le b - a < \int_{r_1}^{N_1} \frac{ds}{\varphi(|s|)}.
$$

Therefore $u'(t_3) < N_1$, and, as t_3 is taken arbitrarily, $u'(t_3) < N_1$, for values of t where $u'(t) > r_1$.

If $t_2 > t_3$, the technique is analogous for $t_4 \in [t_3, t_2]$.

The same conclusion can be achieved if there are $t_2, t_3 \in [a, b]$ such that

$$
u'(t_2) > -r_1 \text{ and } u'(t_3) < -r_1.
$$

Therefore $\|u'\| \le N_1$ and, by similar arguments, it can be proved that $\|v'\| \le N_2$. □

For the reader's convenience we present Schauder's fixed point theorem:

Theorem 2.1 ([269]). *Let Y be a nonempty, closed, bounded and convex subset of a Banach space X, and suppose that $P : Y \to Y$ is a compact operator. Then P has at least one fixed point in Y.*

2.2 Main result

Along this chapter we denote $(a, b) \le (c, d)$ meaning that $a \le c$ and $b \le d$, for $a, b, c, d \in \mathbb{R}$.

Theorem 2.2. *Let $f, h : [a, b] \times \mathbb{R}^4 \to \mathbb{R}$ be continuous functions, and assume that hypothesis (A) holds. If there are coupled lower and upper solutions of (2.1)-(2.2), (α_1, α_2) and (β_1, β_2), respectively, according to Definition 2.1, such that*

$$
(\alpha_1(t), \alpha_2(t)) \le (\beta_1(t), \beta_2(t)), \ \forall t \in [a, b], \tag{2.11}
$$

and f and h verify the Nagumo conditions, then there is at least a pair $(u(t), v(t)) \in \left(C^2[a, b], \mathbb{R}\right)^2$ solution of (2.1)-(2.2) and, moreover,

$$
\alpha_1(t) \le u(t) \le \beta_1(t), \ \alpha_2(t) \le v(t) \le \beta_2(t), \ \forall t \in [a, b],
$$

and

$$
\|u'\| \le N_1 \text{ and } \|v'\| \le N_2,
$$

with N_1 and N_2 given by Lemma 2.1.

Proof. Consider the auxiliary functions $F(t, x, y, z, w) := F$ defined as

- $f\left(t, \beta_1(t), \beta_2(t), \beta_1'(t), \beta_2'(t)\right) - \frac{x-\beta_1(t)}{1+|x-\beta_1(t)|} - \frac{y-\beta_2(t)}{1+|y-\beta_2(t)|}$ if $x > \beta_1(t),\ y > \beta_2(t),$

- $f\left(t, u(t), v(t), u'(t), v'(t)\right) - \frac{y-\beta_2(t)}{1+|y-\beta_2(t)|}$ if $\alpha_1(t) \leq x \leq \beta_1(t),$ $y > \beta_2(t),$

- $f\left(t, \alpha_1(t), \alpha_2(t), \alpha_1'(t), \alpha_2'(t)\right) - \frac{x-\alpha_1(t)}{1+|x-\alpha_1(t)|} + \frac{y-\beta_2(t)}{1+|y-\beta_2(t)|}$ if $x < \alpha_1(t),\ y > \beta_2(t),$

- $f\left(t, \beta_1(t), \beta_2(t), \beta_1'(t), \beta_2'(t)\right) - \frac{x-\beta_1(t)}{1+|x-\beta_1(t)|}$, if $x > \beta_1(t),$ $\alpha_2(t) \leq y \leq \beta_2(t),$

- $f\left(t, u(t), v(t), u'(t), v'(t)\right)$ if $\alpha_1(t) \leq x \leq \beta_1(t),\ \alpha_2(t) \leq y \leq \beta_2(t),$

- $f\left(t, \alpha_1(t), \alpha_2(t), \alpha_1'(t), \alpha_2'(t)\right) - \frac{x-\alpha_1(t)}{1+|x-\alpha_1(t)|}$ if $x < \alpha_1(t),$ $\alpha_2(t) \leq y \leq \beta_2(t),$

- $f\left(t, \beta_1(t), \beta_2(t), \beta_1'(t), \beta_2'(t)\right) - \frac{x-\beta_1(t)}{1+|x-\beta_1(t)|} + \frac{y-\alpha_2(t)}{1+|y-\alpha_2(t)|}$ if $x > \beta_1(t),\ y < \alpha_2(t),$

- $f\left(t, u(t), v(t), u'(t), v'(t)\right) - \frac{y-\alpha_2(t)}{1+|y-\alpha_2(t)|}$ if $\alpha_1(t) \leq x \leq \beta_1(t),$ $y < \alpha_2(t),$

- $f\left(t, \alpha_1(t), \alpha_2(t), \alpha_1'(t), \alpha_2'(t)\right) - \frac{x-\alpha_1(t)}{1+|x-\alpha_1(t)|} - \frac{y-\alpha_2(t)}{1+|y-\alpha_2(t)|}$ if $x < \alpha_1(t),\ y < \alpha_2(t),$

and $H(t, x, y, z, w) := H$ given by

- $h\left(t, \beta_1(t), \beta_2(t), \beta_1'(t), \beta_2'(t)\right) - \frac{x-\beta_1(t)}{1+|x-\beta_1(t)|} - \frac{y-\beta_2(t)}{1+|y-\beta_2(t)|}$ if $x > \beta_1(t),\ y > \beta_2(t),$

- $h\left(t, u(t), v(t), u'(t), v'(t)\right) - \frac{y-\beta_2(t)}{1+|y-\beta_2(t)|}$ if $\alpha_1(t) \leq x \leq \beta_1(t),$ $y > \beta_2(t),$

- $h\left(t, \alpha_1(t), \alpha_2(t), \alpha_1'(t), \alpha_2'(t)\right) - \frac{x-\alpha_1(t)}{1+|x-\alpha_1(t)|} + \frac{y-\beta_2(t)}{1+|y-\beta_2(t)|}$ if $x < \alpha_1(t),\ y > \beta_2(t),$

- $h\left(t, \beta_1(t), \beta_2(t), \beta_1'(t), \beta_2'(t)\right) - \frac{x-\beta_1(t)}{1+|x-\beta_1(t)|}$, if $x > \beta_1(t),$ $\alpha_2(t) \leq y \leq \beta_2(t),$

- $h\left(t, u(t), v(t), u'(t), v'(t)\right)$ if $\alpha_1(t) \leq x \leq \beta_1(t),$ $\alpha_2(t) \leq y \leq \beta_2(t),$

- $h\left(t, \alpha_1(t), \alpha_2(t), \alpha_1'(t), \alpha_2'(t)\right) - \frac{x-\alpha_1(t)}{1+|x-\alpha_1(t)|}$ if $x < \alpha_1(t),$ $\alpha_2(t) \leq y \leq \beta_2(t),$

- $h\left(t, \beta_1(t), \beta_2(t), \beta_1'(t), \beta_2'(t)\right) - \frac{x-\beta_1(t)}{1+|x-\beta_1(t)|} + \frac{y-\alpha_2(t)}{1+|y-\alpha_2(t)|}$ if $x > \beta_1(t),\ y < \alpha_2(t),$

- $h\left(t, u(t), v(t), u'(t), v'(t)\right) - \frac{y-\alpha_2(t)}{1+|y-\alpha_2(t)|}$ if $\alpha_1(t) \leq x \leq \beta_1(t),$ $y < \alpha_2(t),$

- $h\left(t, \alpha_1(t), \alpha_2(t), \alpha_1'(t), \alpha_2'(t)\right) - \frac{x-\alpha_1(t)}{1+|x-\alpha_1(t)|} - \frac{y-\alpha_2(t)}{1+|y-\alpha_2(t)|}$ if $x < \alpha_1(t),\ y < \alpha_2(t),$

and the auxiliary problem

$$\begin{cases} u''(t) = F(t, u(t), v(t), u'(t), v'(t)) \\ v''(t) = H(t, u(t), v(t), u'(t), v'(t)) \\ u(a) = v(a) = 0 \\ u(b) = \delta_1 (b, u(b) + L_1(u, u(b), u'(b))) \\ v(b) = \delta_2 (b, v(b) + L_2(v, v(b), v'(b))), \end{cases} \quad (2.12)$$

where, for each $i = 1, 2$,

$$\delta_i(t, w) = \begin{cases} \beta_i(t), & w > \beta_i(t) \\ w, & \alpha_i(t) \le w \le \beta_i(t) \\ \alpha_i(t), & w < \alpha_i(t). \end{cases} \quad (2.13)$$

Claim 1: *Solutions of problem (2.12) can be written as*

$$u(t) = \frac{t-a}{b-a}\delta_1 (b, u(b) + L_1(u, u(b), u'(b)))$$

$$+ \int_a^b G(t, s)F(s, u(s), v(s), u'(s), v'(s))ds$$

$$v(t) = \frac{t-a}{b-a}\delta_2 (b, v(b) + L_2(v, v(b), v'(b)))$$

$$+ \int_a^b G(t, s)H(s, u(s), v(s), u'(s), v'(s))ds,$$

where

$$G(t, s) = \frac{1}{b-a} \begin{cases} (a-s)(b-t), & a \le t \le s \le b \\ (a-t)(b-s), & a \le s \le t \le b. \end{cases} \quad (2.14)$$

In fact, for the equation $u''(t) = F(t)$, the solution is,

$$u(t) = At + B + \int_a^t (t-s)F(s)ds, \quad (2.15)$$

for some $A, B \in \mathbb{R}$.

By the boundary conditions, it follows that

$$A = \frac{1}{b-a}\delta_1 (b, u(b) + L_1(u, u(b), u'(b))) - \frac{1}{b-a} \int_a^b (b-s)F(s)ds, \quad (2.16)$$

and

$$B = -\frac{a}{b-a}\delta_1 (b, u(b) + L_1(u, u(b), u'(b))) + \frac{a}{b-a} \int_a^b (b-s)F(s)ds.$$

By (2.15) and (2.16), then

$$u(t) = \frac{t-a}{b-a}\delta_1\left(b, u\left(b\right) + L_1(u, u(b), u'(b))\right)$$

$$-\frac{t-a}{b-a}\int_a^b (b-s)F(s)ds + \int_a^t (t-s)F(s)ds$$

$$= \frac{t-a}{b-a}\delta_1\left(b, u\left(b\right) + L_1(u, u(b), u'(b))\right)$$

$$+ \int_a^t \left(\frac{(a-t)(b-s)}{b-a} + t - s\right)F(s)ds + \int_t^b \frac{(a-t)(b-s)}{b-a}F(s)ds$$

$$= \frac{t-a}{b-a}\delta_1\left(b, u\left(b\right) + L_1(u, u(b), u'(b))\right)$$

$$+ \int_a^t \left(\frac{(a-s)(b-t)}{b-a} + t - s\right)F(s)ds + \int_t^b \frac{(a-t)(b-s)}{b-a}F(s)ds$$

$$= \frac{t-a}{b-a}\delta_1\left(b, u\left(b\right) + L_1(u, u(b), u'(b))\right) + \int_a^t G(t,s)F(s)ds,$$

with $G(t,s)$ given by (2.14).

The integral form of $v(t)$ can be achieved by the same arguments.

Claim 2: *Every solution (u, v) of (2.12) satisfies*

$$\|u'\| \le N_1 \text{ and } \|v'\| \le N_2,$$

with N_1 and N_2 given by Lemma 2.1.

This claim is a direct consequence of Lemma 2.1, *as $F(t, x, y, z, w)$ and $H(t, x, y, z, w)$ verify the Nagumo-type conditions.*

Define the operators $T_1 : \left(C^1\left[a, b\right]\right)^2 \to C^1\left[a, b\right]$ and $T_2 : \left(C^1\left[a, b\right]\right)^2 \to C^1\left[a, b\right]$ such that

$$T_1\left(u, v\right)(t) = \frac{t-a}{b-a}\delta_1\left(b, u\left(b\right) + L_1(u, u(b), u'(b))\right)$$

$$+ \int_a^b G(t,s)F(s, u(s), v(s), u'(s), v'(s))ds \qquad (2.17)$$

$$T_2\left(u, v\right)(t) = \frac{t-a}{b-a}\delta_2\left(b, v\left(b\right) + L_2(v, v(b), v'(b))\right)$$

$$+ \int_a^b G(t,s)H(s, u(s), v(s), u'(s), v'(s))ds,$$

where $G(t,s)$ is given by (2.14), and $T : \left(C^1\left[a, b\right]\right)^2 \to \left(C^1\left[a, b\right]\right)^2$ by

$$T\left(u, v\right)(t) = \left(T_1\left(u, v\right)(t), T_2\left(u, v\right)(t)\right). \qquad (2.18)$$

By Claim 1, fixed points of the operator $T := (T_1, T_2)$ are solutions of problem (2.12).

Claim 3: *The operator T, given by (2.18) has a fixed point (u_0, v_0).*

In order to apply Theorem 5.2, we will prove the following steps for operator $T_1 (u, v)$. The proof for the operator $T_2 (u, v)$ is analogous.

(i) $T_1 : \left(C^1 [a, b] \right)^2 \to C^1 [a, b]$ *is well defined.*

The function F is bounded by the Nagumo-type conditions, (2.5), (2.6), and the Green function $G(t, s)$ is continuous in $[a, b]^2$, then the operator $T_1 (u, v)$ is continuous. Moreover, as $\frac{\partial G}{\partial t}(t, s)$ is bounded in $[a, b]^2$ and

$$(T_1 (u, v))' (t) = \frac{1}{b - a} \delta_1 (b, u(b) + L_1(u, u(b), u'(b)))$$
$$+ \int_a^b \frac{\partial G}{\partial t}(t, s) \, F(s, u(s), v(s), u'(s), v'(s)) ds,$$

with

$$\frac{\partial G}{\partial t}(t, s) = \frac{1}{b - a} \begin{cases} s - a, & a \le t \le s \le b, \\ s - b, & a \le s \le t \le b, \end{cases}$$

verifying

$$\left| \frac{\partial G}{\partial t}(t, s) \right| \le 1, \; \forall (t, s) \in [a, b]^2,$$

therefore, $(T_1 (u, v))'$ is continuous on $[a, b]$. So, $T_1 \in C^1 [a, b]$.

(ii) *TB is uniformly bounded, for B a bounded set in $\left(C^1 [a, b] \right)^2$.*

Let B be a bounded set of $\left(C^1 [a, b] \right)^2$. Then there exists $K > 0$ such that

$$\|(u, v)\|_{E^2} = \|u\|_{C^1} + \|v\|_{C^1} \le K, \; \forall (u, v) \in B.$$

By (2.13), and taking into account that F and H are bounded, then there are $M_1, M_2, M_3 > 0$ such that

$$\delta_1 \le \max \{\|\alpha_1\|, \|\beta_1\|\} := M_1,$$

$$\int_a^b \max_{t \in [a, b]} |G(t, s)| \, |F(s, u(s), v(s), u'(s), v'(s))| \, ds \le M_2,$$

$$\int_a^b \max_{t \in [a, b]} \left| \frac{\partial G}{\partial t}(t, s) \right| |F(s, u(s), v(s), u'(s), v'(s))| \, ds \le M_3.$$

Moreover,

$$\|T_1\left(u,v\right)(t)\| = \max_{t\in[a,b]} \left| \frac{t-a}{b-a}\delta_1\left(b,u\left(b\right)+L_1(u,u(b),u'(b))\right) \right.$$

$$\left. + \int_a^b G(t,s)F(s,u(s),v(s),u'(s),v'(s))ds \right|$$

$$\leq \max_{t\in[a,b]} \left| \frac{t-a}{b-a}\right| |\delta_1\left(b,u\left(b\right)+L_1(u,u(b),u'(b)))\right)|$$

$$+ \int_a^b \max_{t\in[a,b]} |G(t,s)| \, |F(s,u(s),v(s),u'(s),v'(s))| \, ds$$

$$\leq M_1 + M_2 < +\infty, \ \forall\,(u,v) \in B,$$

and

$$\|\left(T_1\left(u,v\right)\right)'(t)\| = \max_{t\in[a,b]} \left| \frac{1}{b-a}\delta_1\left(b,u\left(b\right)+L_1(u,u(b),u'(b))\right) \right.$$

$$\left. + \int_a^b \frac{\partial G}{\partial t}(t,s)F(s,u(s),v(s),u'(s),v'(s))ds \right|$$

$$\leq \frac{M_1}{b-a} + \int_a^b \max_{t\in[a,b]} \left|\frac{\partial G}{\partial t}(t,s)\right| |F(s,u(s),v(s),u'(s),v'(s))|ds$$

$$\leq \frac{M_1}{b-a} + M_3 < +\infty, \ \forall\,(u,v) \in B.$$

So, TB is uniformly bounded, for B a bounded set in $\left(C^1\left[a,b\right]\right)^2$.

(iii) *TB is equicontinuous on $\left(C^1\left[a,b\right]\right)^2$.*

Let t_1 and $t_2 \in [a,b]$. Without loss of generality suppose $t_1 \leq t_2$. As $G(t,s)$ is uniformly continuous and F is bounded, then

$$|T_1\left(u,v\right)(t_1) - T_1\left(u,v\right)(t_2)|$$

$$= \left| \frac{(t_1-a)-(t_2-a)}{b-a}\delta_1\left(b,u\left(b\right)+L_1(u,u(b),u'(b))\right) \right.$$

$$\left. + \int_a^b \left[G(t_1,s)-G(t_2,s)\right]F(s,u(s),v(s),u'(s),v'(s))ds \right|$$

$$\leq \left| \frac{t_1-t_2}{b-a}\right| M_1$$

$$+ \left| \int_a^b \left[G(t_1,s)-G(t_2,s)\right] \right| F(s,u(s),v(s),u'(s),v'(s))ds$$

$$\to 0, \ \text{as } t_1 \to t_2,$$

and

$$\left| (T_1(u,v)(t_1))' - (T_1(u,v)(t_2))' \right|$$

$$= \left| \int_a^b \left[\frac{\partial G}{\partial t}(t_1,s) - \frac{\partial G}{\partial t}(t_2,s) \right] F(s,u(s),v(s),u'(s),v'(s)) ds \right|$$

$$\leq \int_a^{t_1} \left| \frac{\partial G}{\partial t}(t_1,s) - \frac{\partial G}{\partial t}(t_2,s) \right| |F(s,u(s),v(s),u'(s),v'(s))| \, ds$$

$$+ \int_{t_1}^{t_2} \left| \frac{\partial G}{\partial t}(t_1,s) - \frac{\partial G}{\partial t}(t_2,s) \right| |F(s,u(s),v(s),u'(s),v'(s))| \, ds$$

$$+ \int_{t_2}^b \left| \frac{\partial G}{\partial t}(t_1,s) - \frac{\partial G}{\partial t}(t_2,s) \right| |F(s,u(s),v(s),u'(s),v'(s))| \, ds.$$

As the function $\frac{\partial G}{\partial t}(t,s)$ has only a jump discontinuity at $t = s$, therefore, as previously, the first and third integrals tend to 0, as $t_1 \to t_2$. For the second integral, as the functions $\frac{\partial G}{\partial t}(t_1,s)$ and $\frac{\partial G}{\partial t}(t_2,s)$ are uniformly continuous, for $s \in [a,t_1[\cup]t_1,b]$ and $s \in [a,t_2[\cup]t_2,b]$, respectively, and

$$\left| \frac{\partial G}{\partial t}(t_1,s) - \frac{\partial G}{\partial t}(t_2,s) \right| |F(s,u(s),v(s),u'(s),v'(s))|$$

is bounded, then

$$\int_{t_1}^{t_2} \left| \frac{\partial G}{\partial t}(t_1,s) - \frac{\partial G}{\partial t}(t_2,s) \right| |F(s,u(s),v(s),u'(s),v'(s))| \, ds \to 0,$$

as $t_1 \to t_2$.

By the Arzèla-Ascoli Theorem $T_1(u,v)$ is compact in $\left(C^1[a,b] \right)^2$.

Following similar arguments with $K_1, K_2, K_3 > 0$ such that

$$\delta_2 \leq \max\{ \|\alpha_2\|, \|\beta_2\| \} := K_1,$$

$$\int_a^b \max_{t \in [a,b]} |G(t,s)| \, |H(s,u(s),v(s),u'(s),v'(s))| \, ds \leq K_2,$$

$$\int_a^b \max_{t \in [a,b]} \left| \frac{\partial G}{\partial t}(t,s) \right| |H(s,u(s),v(s),u'(s),v'(s))| \, ds \leq K_3,$$

it can be shown that $T_2(u,v)$ is compact in $\left(C^1[a,b] \right)^2$, too.

(iv) $TD \subset D$ for some $D \subset \left(C^1[a,b] \right)^2$ a closed and bounded set.

Suppose $D \subset \left(C^1[a,b] \right)^2$ defined by

$$D = \left\{ (u,v) \in \left(C^1[a,b] \right)^2 : \|(u,v)\|_{E^2} \leq 2\rho \right\},$$

where ρ is such that

$$\rho := \max\left\{ M_1 + M_2, \frac{M_1}{b-a} + M_3, K_1 + K_2, \frac{K_1}{b-a} + K_3, N_1, N_2 \right\},$$

with N_1, N_2 given by (2.10).

Arguing as in Claim 3 (ii), it can be shown that

$$\|T_1(u,v)\| \leq M_1 + M_2 \leq \rho,$$

$$\|(T_1(u,v))'\| \leq \frac{M_1}{b-a} + M_3 \leq \rho$$

and, therefore, $\|T_1(u,v)\|_{C^1} \leq \rho$.

Analogously $\|T_2(u,v)\|_{C^1} \leq \rho$ and, so,

$$\|T(u,v)\|_{E^2} = \|(T_1(u,v), T_2(u,v))\|_{E^2}$$
$$= \|T_1(u,v)\|_{C^1} + \|T_2(u,v)\|_{C^1} \leq 2\rho.$$

By Theorem 5.2, the operator T, given by (2.18) has a fixed point (u_0, v_0).

Claim 4: *This fixed point (u_0, v_0) is also a solution of the initial problem (2.1), (2.2), if every solution of (2.12) verifies*

$$\alpha_1(t) \leq u_0(t) \leq \beta_1(t), \quad \alpha_2(t) \leq v_0(t) \leq \beta_2(t), \quad \forall t \in [a,b], \qquad (2.19)$$
$$\alpha_1(b) \leq u_0(b) + L_1(u_0, u_0(b), u_0'(b)) \leq \beta_1(b), \qquad (2.20)$$
$$\alpha_2(b) \leq v_0(b) + L_2(v_0, v_0(b), v_0'(b)) \leq \beta_2(b). \qquad (2.21)$$

Let (u_0, v_0) be a fixed point of T, that is (u_0, v_0) is a fixed point of T_1 and T_2.

By Claim 1, (u_0, v_0) is solution of problem (2.12).

In the following we will prove the estimations for u_0, as for v_0 the procedure is analogous.

Suppose, by contradiction, that the first inequality of (2.19) is not true. So, there exists $t \in [a,b]$ such that $\alpha_1(t) > u_0(t)$ and it can be defined

$$\max_{t \in [a,b]} (\alpha_1(t) - u_0(t)) := \alpha_1(t_0) - u_0(t_0) > 0. \qquad (2.22)$$

Remark that, by (2.12), Definition 2.1 and (2.13), $t_0 \neq a$, as $\alpha_1(a) - u_0(a) \leq 0$, and $t_0 \neq b$, because

$$\alpha_1(b) - u_0(b) = \alpha_1(b) - \delta_1(b, u_0(b) + L_1(u_0, u_0(b), u_0'(b))) \leq 0.$$

Then $t_0 \in \,]a, b[$,

$$\alpha_1'(t_0) - u_0'(t_0) = 0 \text{ and } \alpha_1''(t_0) - u_0''(t_0) \leq 0. \qquad (2.23)$$

There are three possibilities for the value of $v_0(t_0)$:

- If $v_0(t_0) > \beta_2(t_0)$, then, by (2.12) and Definition 2.1, the following contradiction with (2.23) is obtained

$$u_0''(t_0) = F(t_0, u_0(t_0), v_0(t_0), u_0'(t_0), v_0'(t_0))$$
$$= f(t_0, \alpha_1(t_0), \alpha_2(t_0), \alpha_1'(t_0), \alpha_2'(t_0)) - \frac{u_0(t_0) - \alpha_1(t_0)}{1 + |u_0(t_0) - \alpha_1(t_0)|}$$
$$+ \frac{v_0(t_0) - \beta_2(t_0)}{1 + |v_0(t_0) - \beta_2(t_0)|}$$
$$> f(t_0, \alpha_1(t_0), \alpha_2(t_0), \alpha_1'(t_0), \alpha_2'(t_0)) \geq \alpha_1''(t_0).$$

- If $\alpha_2(t_0) \leq v_0(t_0) \leq \beta_2(t_0)$, then

$$u_0''(t_0) = F(t_0, u_0(t_0), v_0(t_0), u_0'(t_0), v_0'(t_0))$$
$$= f(t_0, \alpha_1(t_0), \alpha_2(t_0), \alpha_1'(t_0), \alpha_2'(t_0)) - \frac{u_0(t_0) - \alpha_1(t_0)}{1 + |u_0(t_0) - \alpha_1(t_0)|}$$
$$> f(t_0, \alpha_1(t_0), \alpha_2(t_0), \alpha_1'(t_0), \alpha_2'(t_0)) \geq \alpha_1''(t_0).$$

- If $\alpha_2(t_0) < v_0(t_0)$, the contradiction is

$$u_0''(t_0) = F(t_0, u_0(t_0), v_0(t_0), u_0'(t_0), v_0'(t_0))$$
$$= f(t_0, \alpha_1(t_0), \alpha_2(t_0), \alpha_1'(t_0), \alpha_2'(t_0)) - \frac{u_0(t_0) - \alpha_1(t_0)}{1 + |u_0(t_0) - \alpha_1(t_0)|}$$
$$+ \frac{v_0(t_0) - \alpha_2(t_0)}{1 + |v_0(t_0) - \alpha_2(t_0)|}$$
$$> f(t_0, \alpha_1(t_0), \alpha_2(t_0), \alpha_1'(t_0), \alpha_2'(t_0)) \geq \alpha_1''(t_0).$$

Therefore $\alpha_1(t) \leq u_0(t)$, $\forall t \in [a, b]$.

By a similar technique it can be shown that $u_0(t) \leq \beta_1(t)$, $\forall t \in [a, b]$, and so,

$$\alpha_1(t) \leq u_0(t) \leq \beta_1(t), \forall t \in [a, b]. \tag{2.24}$$

Assume now that, to prove the first inequality of (2.20),

$$\alpha_1(b) > u_0(b) + L_1(u_0, u_0(b), u_0'(b)). \tag{2.25}$$

Then, by (2.12) and (2.13),

$$u_0(b) = \delta_1(b, u_0(b) + L_1(u_0, u_0(b), u_0'(b))) = \alpha_1(b),$$

and

$$u_0'(b) \leq \alpha_1'(b).$$

By (2.25), the assumption (A) and Definition 2.1, we have the contradiction

$$0 > L_1(u_0, u_0(b), u_0'(b)) + u_0(b) - \alpha_1(b)$$
$$\geq L_1(u_0, u_0(b), u_0'(b)) \geq L_1(\alpha_1, \alpha_1(b), \alpha_1'(b)) \geq 0.$$

Then $\alpha_1(b) \leq u_0(b) + L_1(u_0, u_0(b), u_0'(b))$.

To prove the second inequality of (2.20), assume that

$$u_0(b) + L_1(u_0, u_0(b), u_0'(b)) > \beta_1(b). \qquad (2.26)$$

Then, by (2.12) and (2.13),

$$u_0(b) = \delta_1(b, u_0(b) + L_1(u_0, u_0(b), u_0'(b))) = \beta_1(b). \qquad (2.27)$$

By (2.26), (2.27) and Definition 2.1, we have a contradiction with (2.26). So, $u_0(b) + L_1(u_0, u_0(b), u_0'(b)) \leq \beta_1(b)$, and, therefore, (2.20) holds.

To prove (2.21) the technique is analogous.

So, the fixed point (u_0, v_0) of T, solution of problem (2.12), is a solution of problem (2.1), (2.2), too. $\qquad \square$

2.3   Example

Consider the boundary value problem composed by the second-order coupled systems with full nonlinearities

$$\begin{cases} u''(t) = -u'(t)\, v(t) + \arctan(u(t)\, v'(t)) + t, \\[2mm] v''(t) = t^2 \left[-e^{-|u'(t)|}\, v(t) + u(t)\, (v'(t) - 2) \right] \end{cases} \qquad (2.28)$$

with $t \in [0, 1]$, and the functional boundary conditions

$$\begin{cases} u(0) = v(0) = 0 \\[3mm] u(1) = \max_{t \in [0,1]} u(t) - (u'(1))^2 \\[3mm] v(1) = 2 \int_0^1 v(s)\,ds - \dfrac{(v'(1))^3}{8}. \end{cases} \qquad (2.29)$$

This problem is a particular case of system (2.1)-(2.2) with

$$f(t, x, y, z, w) = -z\, y + \arctan(x\, w) + t,$$

$$h(t, x, y, z, w) = t^2 \left[-e^{-|z|}\, y + x\,(w - 2) \right], \qquad (2.30)$$

continuous functions, $t \in [0, 1]$, $a = 0$, $b = 1$, and

$$L_1(w, x, y) = x - \max_{t \in [0,1]} w(t) + y^2,$$

$$L_2(w, x, y) = x - 2 \int_0^1 w(s)\,ds + \frac{y^3}{8}. \qquad (2.31)$$

Remark that, L_1 and L_2 are continuous functions, verifying (A). The functions given by

$$(\alpha_1(t), \alpha_2(t)) = (-1, -1) \text{ and } (\beta_1(t), \beta_2(t)) = (3 + t, 2t + 2)$$

are, respectively, lower and upper solutions of problem (2.28)-(2.29), satisfying (2.11), as, by Definition 2.1, we have

$$\alpha_1''(t) = 0 \le f(t, -1, -1, 0, 0) = t, \ t \in [0, 1]$$
$$\alpha_2''(t) = 0 \le h(t, -1, -1, 0, 0) = 3t^2, \ t \in [0, 1]$$
$$L_1(\alpha_1, \alpha_1(1), \alpha_1'(1)) = 0 \ge 0$$
$$L_2(\alpha_2, \alpha_2(b), \alpha_2'(b)) = \frac{7}{8} \ge 0,$$

and

$$\beta_1(t)''(t) = 0 \ge f(t, 3 + t, 2t + 2, 1, 2)$$
$$= -(2t + 2) + \arctan(6t + 2) + t, \ t \in [0, 1]$$
$$\beta_2''(t) = 0 \ge h(t, 3 + t, 2t + 2, 1, 2) = -\frac{2t^3 + 2t^2}{e}, \ t \in [0, 1]$$
$$L_1(\beta_1, \beta_1(1), \beta_1'(1)) = 0 \le 0$$
$$L_2(\beta_2, \beta_2(b), \beta_2'(b)) = -1 \le 0.$$

Furthermore, the functions f and h, given by (2.30), satisfy Nagumo conditions relative to the intervals $[-1, 2 + t]$ and $[-1, 2t + 1]$, for $t \in [0, 1]$, with $r_1 = r_2 = 4$,

$$|f(t, x, y, z, w)| \le 3|z| + \frac{\pi}{2} + 1 := \varphi(|z|),$$

$$|h(t, x, y, z, w)| \le 3(1 + |w - 2|) := \psi(|w|),$$

and

$$\int_4^{N_1} \frac{ds}{\varphi(|s|)} = \int_4^{N_1} \frac{1}{3s + \frac{\pi}{2} + 1} ds > 1,$$

for $N_1 \ge 100$ and

$$\int_4^{N_2} \frac{ds}{\psi(|s|)} = \int_4^{N_2} \frac{1}{3(1 + |s - 2|) ds} > 1,$$

for $N_2 \ge 80$.

Then, by Theorem 2.2, there is at least a pair $(u(t), v(t)) \in (C^2[0,1], \mathbb{R})^2$ solution of (2.28)-(2.29) and, moreover,

$$-1 \le u(t) \le 2 + t, \quad -1 \le v(t) \le 2t + 1, \ \forall t \in [0,1],$$
$$\|u'\| \le N_1 \text{ and } \|v'\| \le N_2.$$

2.4 Coupled mass-spring system

Consider the mass-spring system composed by two springs with constants of proportionality k_1 and k_2, and two weights of mass m_1 and m_2. The mass m_1 is suspended vertically from a fixed support by a spring with constant k_1 and the mass m_2 is attached to the first weight by a spring with constant k_2. The system described is illustrated in Figure 2.1.

Let us call $u(t)$ and $v(t)$ the displacements of the weights of mass m_1 and m_2, respectively, in relation to their respective equilibrium positions. Thus, at time t, the position of the displacement of the mass m_1 is $u(t)$ and the displacement of mass m_2 is $v(t)$.

For simplicity we consider $t \in [0,1]$, and, therefore, $u(0)$ and $u(1)$ are the initial and final displacements of mass m_1, and $v(0)$ and $v(1)$ are the similar displacements of mass m_2.

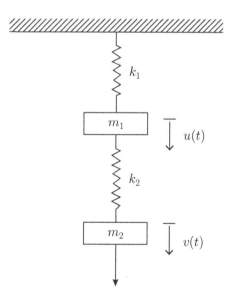

Fig. 2.1 The coupled springs.

As it can be seen in [89], the above system is modelled by the forced second-order nonlinear coupled systems and with friction

$$\begin{cases} m_1 u''(t) = -\delta_1 u'(t) - \kappa_1 u(t) + \mu_1 (u(t))^3 - \kappa_2 (u(t) - v(t)) \\ \qquad\qquad + \mu_2 (u(t) - v(t))^3 + F_1 \cos(\omega_1 t), \\ \\ m_2 v''(t) = -\delta_2 v'(t) - \kappa_2 (v(t) - u(t)) + \mu_2 (v(t) - u(t))^3 \\ \qquad\qquad + F_2 \cos(\omega_2 t) \end{cases} \tag{2.32}$$

where $t \in [0, 1]$,

- δ_1, δ_2 are the damping coefficients;
- μ_1, μ_2 are the coefficients of the nonlinear terms of each system equation;
- $\kappa_2(u(t) - v(t)) + \mu_2(u(t) - v(t))^3$ and $\kappa_2(v(t) - u(t)) + \mu_2(v(t) - u(t))^3$ are the nonlinear restoring forces;
- F_1, F_2 are the forcing amplitudes of the sinusoidal forces $F_1 \cos(\omega_1 t)$ and $F_2 \cos(\omega_2 t)$, where ω_1, ω_2 are the forcing frequencies.

In this work we add to the system the functional boundary conditions

$$\begin{cases} u(0) = v(0) = 0 \\ u(1) = \max_{t \in [0,1]} u(t) + 2u'(1) \\ v(1) = \max_{t \in [0,1]} v(t) + 2\left(v'(1)\right)^3. \end{cases} \tag{2.33}$$

The functional conditions (2.33) can have a physical meaning such as, for example, the first one can be seen as the displacement of mass 1 at the final moment given by the sum of the maximum displacement in this period of time, with the double of the velocity of the displacement at the end point.

Clearly, the above model (2.32), (2.33) is a particular case of system (2.1)-(2.2) with

$$f(t, x, y, z, w) = \frac{1}{m_1} \left[-\delta_1 z - \kappa_1 x + \mu_1 x^3 - \kappa_2 (x - y) + \mu_2 (x - y)^3 \right.$$
$$\left. + F_1 \cos(\omega_1 t) \right],$$

$$h(t, x, y, z, w) = \frac{1}{m_2} \left[-\delta_2 w - \kappa_2 (y - x) + \mu_2 (y - x)^3 + F_2 \cos(\omega_2 t) \right].$$

$$\tag{2.34}$$

These functions are continuous in $[0, 1] \times \mathbb{R}^4$, and

$$\begin{aligned} L_1(w, x, y) &= x - \max_{t \in [0,1]} w(t) - 2y, \\ L_2(w, x, y) &= x - \max_{t \in [0,1]} w(t) - 2y^3 \end{aligned} \tag{2.35}$$

verify (A).

The functions given by

$$(\alpha_1(t),\, \alpha_2(t)) = (-t,\, -t) \quad \text{and} \quad (\beta_1(t),\, \beta_2(t)) = (t,\, t)$$

are, respectively, lower and upper solutions of problem (2.32)-(2.33), satisfying (2.11), for every positive m_1, m_2, non negative δ_1, δ_2, F_1, F_2, ω_1, ω_2, and any real κ_1, κ_2, μ_1, μ_2, such that

$$\begin{aligned} F_1 &\leq \delta_1, \\ F_2 &\leq \delta_2, \\ \mu_1 &\leq 0. \end{aligned} \tag{2.36}$$

Indeed, Definition 2.1 holds because,

$$\alpha_1''(t) = 0 \leq \frac{1}{m_1}\left(\delta_1 - F_1\right)$$

$$\leq \frac{1}{m_1}\left[\delta_1 - \mu_1 t^3 + F_1 \cos(\omega_1 t)\right], \ \forall t \in [0,\,1]$$

$$\leq \frac{1}{m_1}\left[\delta_1 + \kappa_1 t - \mu_1 t^3 + F_1 \cos(\omega_1 t)\right], \ \forall t \in [0,\,1]$$

$$\alpha_2''(t) = 0 \leq \frac{1}{m_2}\left(\delta_2 - F_2\right) \leq \frac{1}{m_2}\left[\delta_2 + F_2 \cos(\omega_2 t)\right], \ \forall t \in [0,\,1]$$

$$L_1(\alpha_1, \alpha_1(1), \alpha_1'(1)) = \alpha_1(1) - \max_{t \in [0,1]} \alpha_1(t) - 2\alpha_1'(1) = 1 \geq 0$$

$$L_2(\alpha_2, \alpha_2(b), \alpha_2'(b)) = \alpha_2(1) - \max_{t \in [0,1]} \alpha_2(t) - 2(\alpha_2'(1))^3 = 1 \geq 0,$$

and

$$\beta_1''(t) = 0 \geq \frac{1}{m_1}\left(-\delta_1 + F_1\right)$$

$$\geq \frac{1}{m_1}\left[-\delta_1 - \kappa_1 t + \mu_1 t^3 + F_1 \cos(\omega_1 t)\right], \ \forall t \in [0,\,1]$$

$$\beta_2''(t) = 0 \geq \frac{1}{m_2}\left(-\delta_2 + F_2\right) \geq \frac{1}{m_2}\left[-\delta_2 + F_2 \cos(\omega_2 t)\right], \ \forall t \in [0,\,1]$$

$$L_1(\beta_1, \beta_1(1), \beta_1'(1)) = \beta_1(1) - \max_{t \in [0,1]} \beta_1(t) - 2\beta_1'(1) = -2 \leq 0$$

$$L_2(\beta_2, \beta_2(b), \beta_2'(b)) = \beta_2(1) - \max_{t \in [0,1]} \beta_2(t) - 2(\beta_2'(1))^3 = -2 \leq 0.$$

Furthermore, f and h, given by (2.34), verify Nagumo-type conditions relative to the interval $[-t, t]$, for $t \in [0,\,1]$, with $r_1 = r_2 = 1$, for

$$-t \leq x \leq t, \ -t \leq y \leq t, \forall t \in [0,1],$$

$$|f(t, x, y, z, w)| \leq \frac{1}{m_1}\left(\delta_1\,|z| + \kappa_1 + |\mu_1| + 2\kappa_2 + 8\,|\mu_2| + F_1\right) := \varphi\,(|z|),$$

$$|h(t, x, y, z, w)| \leq \frac{1}{m_2}\left(\delta_2\,|w| + 2\kappa_2 + 8\,|\mu_2| + F_2\right) := \psi\,(|w|),$$

and for N_1 and N_2 positive, and large enough such that

$$\int_1^{N_1} \frac{ds}{\varphi(s)} = \int_1^{N_1} \left(\frac{m_1}{\delta_1 s + \kappa_1 + |\mu_1| + 2\kappa_2 + 8|\mu_2| + F_1} \right) ds > 1,$$

and

$$\int_1^{N_2} \frac{ds}{\psi(s)} = \int_1^{N_2} \left(\frac{m_2}{\delta_2 s + 2\kappa_2 + 8|\mu_2| + F_2} \right) ds > 1.$$

So, by Theorem 2.2, there is a solution $(u(t), v(t))$ of the mass-spring system (2.32), (2.33), for the values of coefficients verifying (2.36), such that

$$-t \le u(t) \le t,$$
$$-t \le v(t) \le t, \forall t \in [0, 1],$$

that is, both with values in the strip given by Figure 2.2 and

$$\|u'\| \le N_1 \text{ and } \|v'\| \le N_2.$$

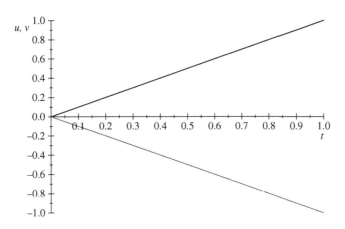

Fig. 2.2   The functions $u(t)$ and $v(t)$ lie in the strip.

Conclusions and open issues

Both chapters of this first Part contain sufficient conditions on the several nonlinearities to have results of existence and localization of a solution for coupled systems: in Chapter 1 the existence and localization of a solution are proved via the well-known Guo-Krasnosel'ski theorem on an adequate compression-expansion cone. Chapter 2 combines the lower and upper solutions method with Schauder's fixed point theorem, and the solution is localized in a branch bounded from below by the lower solution, and from above by the upper solution. Remark that in the first chapter the kernel functions are non-negative, which is not the case in the second chapter.

The arguments used stress some new open questions, that remain to be answered by future work, such as, for instance:

- How to deal with the resonant case, that is, where the associated linear operator is not invertible, and therefore the Green functions can not be defined?
- What are the sufficient conditions for the solvability of coupled systems with n fully differential equations, that is, with dependence on all derivatives?
- Is it possible to consider functional boundary or nonlocal boundary conditions, with functional dependence on all derivatives?
- Can the monotonicity of the functional conditions be overcomed or replaced by other ones?

PART II

Coupled systems on unbounded domains

Introduction

The theory and techniques used in the previous chapter can not be applied to systems defined on unbounded domains, mainly due to the lack of compactness of the associated operator, cases in which the literature for differential equation systems is scarce. Compared to boundary value problems in bounded intervals, we can say that in the literature there is a lack of results and publications that guarantee the existence of solutions for nonlinear coupled systems on unbounded intervals. To the best of our knowledge, so far there has been no work on nonlinear coupled systems with homoclinic or heteroclinic solutions, which will be addressed in this Part.

The first appearance of boundary value problems on unbounded intervals is related to A. Kneser [147], where it was discussed monotone solutions and their derivatives on $[0, \infty)$ for the second-order ordinary differential equation

$$\frac{d^2 y}{dx^2} = y(x).$$

Assuming some conditions, Krasnosel'skii in [152], was one of the first authors to treat fixed-point problems in Banach spaces on unbounded domains, where it was considered the translation operator and periodic solutions of differential equations in Banach space. In particular he studied

$$\frac{dx}{dt} = A(t)x + \phi(t, x),$$

where $A(t)$ is a linear continuous operator that depends continuously (with respect to the norm of the operators) on t and where $\phi(t, x)$ is completely continuous in the sense that it is continuous with respect to the variables $t \in (-\infty, +\infty)$, x belongs to a Banach space E and maps the Cartesian product of each finite interval and an arbitrary ball in the space E into a compact subset of E.

In [19] and the references therein, we can find historical remarks about BVPs at infinite intervals, from the early 1970s, as well as the main techniques used to address these problems. Recently, in [192], Minhós and Carrasco addressed several nonlinear higher order problems in unbounded domains with applications to various real phenomena.

In this Part we study problems composed of nonlinear second-order systems with complete nonlinearities and with boundary conditions at half-line or real line. More precisely, we seek to guarantee the existence and also the localization of solutions on unbounded domain using different approaches, depending on the problem being studied and on the boundary conditions.

The research methodology followed in this Part is based essentially in [193, 196, 199] and we underline the main features:

- Application of a *Nagumo-type growth* condition on nonlinearities and the concept of *equiconvergence* at ∞, to recover the compactness of the associated operators;
- Choice of an *adequate functional* context and the consideration of *asymptotic conditions* and *growth assumptions* on the nonlinearities;
- Based on *phi-Laplacians* and on the consideration of *growth and asymptotic conditions for homeomorphisms* and nonlinearities.

This second Part consists of three chapters which cover the existence and location of coupled systems on unbounded domains, where:

- In the *third chapter* we consider existence and localization of solutions for nonlinear second-order coupled systems with boundary conditions on the semi-infinite interval $[0, \infty)$;
- In the *fourth chapter*, we consider the existence and location of homoclinic solutions for nonlinear second-order coupled systems on real line;
- In the *fifth chapter* we apply previous arguments to the existence of heteroclinic solutions for nonlinear second-order coupled systems on $(-\infty, +\infty)$.

Chapter 3

Second-order coupled systems on the half-line

In the present chapter we study the second-order system defined in a non-compact domain,

$$\begin{cases} u''(t) = f(t, u(t), v(t), u'(t), v'(t)), \ t \in [0, +\infty[, \\ v''(t) = h(t, u(t), v(t), u'(t), v'(t)), \end{cases} \tag{3.1}$$

with $f, h : [0, +\infty[\times \mathbb{R}^4 \to \mathbb{R}$ be L^1-Carathéodory functions, together with the boundary conditions

$$\begin{cases} u(0) = A_1, \ v(0) = A_2, \\ u'(+\infty) = B_1, \ v'(+\infty) = B_2, \end{cases} \tag{3.2}$$

with $A_1, A_2, B_1, B_2 \in \mathbb{R}$,

$$u'(+\infty) := \lim_{t \to +\infty} u'(t) \text{ and } v'(+\infty) := \lim_{t \to +\infty} v'(t).$$

The study of second order differential equations on the half-line, can be seen in [28, 39, 54, 94, 99, 108, 162, 163, 185, 200, 254, 260, 261].

These types of equations have many applications in physics, biology, mechanics and among other areas. Examples of models of specific phenomena can be found on dynamical rotations, gravitational and electrostatic interactions [73], damped and undamped oscillations of a rigid pendulum, circuit analogue and spring mass-systems [275], motion of a piston inside a cylinder and micro-electro-mechanical systems [250], cross diffusion epidemiology and thermographic tumor detection [233].

However, coupled systems where the nonlinearities may depend on different and all of the unknown functions are scarce in the literature.

In [173] it is studied the BVP for second-order singular differential system on the whole line with impulse effects, i.e., consisting of the differential system

$$[\phi_p(\rho(t)x'(t))]' = f(t, x(t), y(t)), \quad \text{a.e. } t \in \mathbb{R},$$
$$[\phi_q(\varrho(t)y'(t))]' = g(t, x(t), y(t)), \quad \text{a.e. } t \in \mathbb{R}$$

subjected to the boundary conditions

$$\lim_{t \to \pm\infty} x(s) = 0,$$
$$\lim_{t \to \pm\infty} y(s) = 0$$

and the impulse effects

$$\Delta x(t_k) = I_k(t_k, x(t_k), y(t_k)), \quad k \in \mathbb{Z},$$
$$\Delta y(t_k) = J_k(t_k, x(t_k), y(t_k)), \quad k \in \mathbb{Z},$$

where

(a) $\rho, \varrho \in C^0(\mathbb{R}, [0, \infty))$, $\rho(t), \varrho(t) > 0$ for all $t \in \mathbb{R}$ with $\int_{-\infty}^{+\infty} \frac{ds}{\rho(s)} < +\infty$
   and $\int_{-\infty}^{+\infty} \frac{ds}{\varrho(s)} < +\infty$,
(b) $\phi_p(x) = x|x|^{p-2}$, $\phi_q(x) = x|x|^{q-2}$ with $p > 1$ and $q > 1$ are p-Laplacian
   and q-Laplacian operators;
(c) f, g on $\mathbb{R}^3$ are Carathéodory functions;
(d) t_k is an increasing sequence, $k \in \mathbb{Z}$, $\cdots < t_k < t_{k+1} < t_{k+2} < \cdots$ with
   $\lim_{k \to -\infty} t_k = -\infty$ and $\lim_{k \to +\infty} t_k = +\infty$, $\Delta x(t_k) = u(t_k^+) - x(t_k^-)$
   and $\Delta y(t_k) = y(t_k^+) - y(t_k^-)$ and $\mathbb{Z}$ is the set of all integers;
(e) $\{I_k\}, \{J_k\}$, with $I_k, J_k : \mathbb{R}^3 \to \mathbb{R}$ are Carathéodory sequences.

In [211], the authors study the existence of nontrivial solutions to the boundary value problem

$$\begin{cases} -u'' + cu' + \lambda u = f(x, u), & -\infty < x < +\infty, \\ u(-\infty) = u(+\infty) = 0, \end{cases}$$

and to the system

$$\begin{cases} u'' + c_1 u' + \lambda_1 u = f(x, u, v), & -\infty < x < +\infty, \\ v'' + c_2 v' + \lambda_2 v = g(x, u, v), & -\infty < x < +\infty, \\ u(-\infty) = u(+\infty) = 0, \ v(-\infty) = v(+\infty) = 0, \end{cases}$$

where $c, c_1, c_2, \lambda, \lambda_1, \lambda_2$ are real positive constants and the nonlinearities f and g satisfy suitable conditions.

Motivated by these works, our method follows arguments applied in [199], for differential equations in the half-line, and in [205], for coupled systems. Note that, our technique is based on [197], where it was the first time where coupled systems of differential equations are considered, namely with first derivative dependence on both unknown variables in unbounded domains. To be more precise, we use lower and upper solutions method combined with a Nagumo type growth condition. The equiconvergence at infinity plays a key role to recover the compactness of the correspondent operators.

3.1 Definitions and preliminary results

Define the space
$$X = \left\{ x : x \in C^1([0, +\infty[) : \lim_{t \to +\infty} \frac{|x(t)|}{1+t} \in \mathbb{R}, \ \lim_{t \to +\infty} |x'(t)| \in \mathbb{R} \right\},$$
and the norm $\|w\|_X = \max\{\|w\|_0, \|w'\|_1\}$, where
$$\|\Upsilon\|_0 := \sup_{t \in [0, +\infty[} \frac{|\Upsilon(t)|}{1+t} \quad \text{and} \quad \|\Upsilon\|_1 := \sup_{t \in [0, +\infty[} |\Upsilon(t)|.$$

Denoting $E := X \times X$ with the norm
$$\begin{aligned}
\|(u, v)\|_E &= \max\{\|u\|_X, \|v\|_X\} \\
&= \max\{\|u\|_0, \|u'\|_1, \|v\|_0, \|v'\|_1\}.
\end{aligned}$$

So, $(E, \|.\|_E)$ is a Banach space.

L^1-Carathéodory functions are considered in the space X:

Definition 3.1. A function $g : [0, +\infty[\times \mathbb{R}^4 \to \mathbb{R}$ is a L^1-Carathéodory if

(i) for each $(x, y, z, w) \in \mathbb{R}^4$, $t \mapsto g(t, x, y, z, w)$ is measurable on $[0, +\infty[$;

(ii) for almost every $t \in [0, +\infty[$, $(x, y, z, w) \mapsto g(t, x, y, z, w)$ is continuous on $\mathbb{R}^4$;

(iii) for each $\rho > 0$, there exists a positive function $\phi_\rho \in L^1([0, +\infty[)$ such that, for $(x, y, z, w) \in \mathbb{R}^4$
$$\sup_{t \in [0, +\infty[} \left\{ \frac{|x|}{1+t}, \frac{|y|}{1+t}, |z|, |w| \right\} < \rho, \tag{3.3}$$
one has
$$|g(t, x, y, z, w)| \leq \phi_\rho(t), \quad \text{a.e. } t \in [0, +\infty[.$$

Take $\alpha_1, \beta_1, \alpha_2, \beta_2 \in C([0, +\infty[)$ such that, for $t \in [0, +\infty[$, $\alpha_1(t) \leq \beta_1(t)$, $\alpha_2(t) \leq \beta_2(t)$ and define the following set,

$$S = \{(t, x, y, z, w) \in [0, +\infty[\times \mathbb{R}^4 : \alpha_1(t) \leq x \leq \beta_1(t),\ \alpha_2(t) \leq y \leq \beta_2(t)\}.$$

Definition 3.2. The functions $f, h : S \to \mathbb{R}$, satisfy a Nagumo-type growth condition in S, if for some positive continuous functions ϕ, φ, l_1, l_2 and some $\varepsilon > 1$, $R_1, R_2 > 0$, such that

$$\sup_{t \in [0, +\infty[} \phi(t)(1 + t)^\varepsilon < R_1, \qquad \int_0^{+\infty} \frac{s}{l_1(s)} ds = +\infty, \qquad (3.4)$$

$$\sup_{t \in [0, +\infty[} \varphi(t)(1 + t)^\varepsilon < R_2, \qquad \int_0^{+\infty} \frac{s}{l_2(s)} ds = +\infty, \qquad (3.5)$$

$$|f(t, x, y, z, w)| \leq \phi(t) l_1(|z|) \ \forall (t, x, y, z, w) \in S \qquad (3.6)$$

and

$$|h(t, x, y, z, w)| \leq \varphi(t) l_2(|w|) \ \forall (t, x, y, z, w) \in S. \qquad (3.7)$$

Lemma 3.1. *Let $f, h : [0, +\infty[\times \mathbb{R}^4 \to \mathbb{R}$ be L^1-Carathéodory functions, where f and h satisfy (3.6), (3.7), and $\varepsilon > 1$. Then for every solution $(u, v) \in (C^2[0, +\infty[)^2 \cap E$, verifying the inequalities,*

$$\alpha_1(t) \leq u(t) \leq \beta_1(t), \ \alpha_2(t) \leq v(t) \leq \beta_2(t), \ \forall t \in [0, +\infty[, \qquad (3.8)$$

there are $N_1, N_2 > 0$ such that

$$\|u'\|_1 \leq N_1 \ and \ \|v'\|_1 \leq N_2. \qquad (3.9)$$

Proof. Let $(u(t), v(t))$ be a solution of (3.1), (3.2) and consider $r > 0$ such that $r > \max\{|B_1|, |B_2|\}$.

If $|u'(t)| \leq r$, $\forall t \in [0, +\infty[$, then taking $N_1 \geq r$ the proof would be concluded.

As, by (3.2), $|u'(t)| > r$, $\forall t \in [0, +\infty[$, can not happen. Then assume that there is $t \in [0, +\infty[$ such that $u'(t) > r$. Therefore, there are $t_0, t_1 \in [0, +\infty[$ such that $t_0 < t_1$, $u'(t_1) = r$ and, for $t \in [t_0, t_1[$, we have $u'(t) > r$.

Take $N_i > r$, $i = 1, 2$, such that

$$\int_r^{N_i} \frac{s}{l_i(s)} ds > R_i \max \left\{ \begin{array}{l} M_i + \sup_{t \in [0, +\infty[} \frac{\beta_i(t)}{1+t} \frac{\varepsilon}{\varepsilon - 1}, \\ M_i - \inf_{t \in [0, +\infty[} \frac{\alpha_i(t)}{1+t} \frac{\varepsilon}{\varepsilon - 1} \end{array} \right\},$$

for $i = 1, 2$, R_i given by (3.4) and (3.5), and

$$M_i := \sup_{t \in [0, +\infty[} \frac{\beta_i(t)}{(1+t)^\varepsilon} - \inf_{t \in [0, +\infty[} \frac{\alpha_i(t)}{(1+t)^\varepsilon}.$$

Thus, by a convenient change of variable and (3.6),

$$\int_{u'(t_0)}^{u'(t_1)} \frac{s}{l_1(s)} ds = \int_{t_0}^{t_1} \frac{u'(s)}{l_1(u'(s))} u''(s) ds$$

$$\leq \int_{t_0}^{t_1} \frac{u'(s)}{l_1(u'(s))} |f(u(s), v(s), u'(s), v'(s))| ds \leq \int_{t_0}^{t_1} \phi(s) u'(s) ds$$

$$\leq \int_{t_0}^{t_1} \frac{R_1 u'(s)}{(1+s)^\varepsilon} ds = R_1 \left[\int_{t_0}^{t_1} \left(\frac{u(s)}{(1+s)^\varepsilon} \right)' + \frac{\varepsilon u(s)}{(1+s)^{1+\varepsilon}} \right] ds$$

$$\leq R_1 \left(\sup_{t \in [0, +\infty[} \frac{\beta_1(t)}{(1+t)^\varepsilon} - \inf_{t \in [0, +\infty[} \frac{\alpha_1(t)}{(1+t)^\varepsilon} + \int_{t_0}^{t_1} \frac{\varepsilon u(s)}{(1+s)^{1+\varepsilon}} ds \right)$$

$$= R_1 \left(M_1 + \int_{t_0}^{t_1} \frac{\varepsilon u(s)}{(1+s)^{1+\varepsilon}} ds \right) \leq R_1 \left(M_1 + \int_{t_0}^{t_1} \frac{\beta_1(s)}{(1+s)^{1+\varepsilon}} \varepsilon ds \right)$$

$$\leq R_1 \left(M_1 + \sup_{t \in [0, +\infty[} \frac{\beta_1(t)}{(1+t)} \int_0^{+\infty} \frac{\varepsilon}{(1+s)^\varepsilon} ds \right) \leq \int_r^{N_1} \frac{s}{l_1(s)} ds.$$

Since we can take arbitrarily $t_0 \in [0, +\infty[$, such that $u'(t_0) > r$, therefore $u'(t_0) < N_1$.

If $u'(t) < -r$ the technique is analogous. The same conclusion can be achieved if there are $t_{-1}, t_2 \in [0, +\infty[$ such that $t_{-1} < t_2$, $u'(t_2) = -r$ and $u'(t) < -r$, $\forall t \in [t_{-1}, t_2[$.

Therefore $\|u'\| \leq N_1$. By similar arguments, it can be proved that $\|v'\| \leq N_2$. □

Lemma 3.2. *Let $f, h : [0, +\infty[\times \mathbb{R}^4 \to \mathbb{R}$ be L^1-Carathéodory functions. Then the coupled system*

$$\begin{cases} u''(t) = f(t, u(t), v(t), u'(t), v'(t)), \ t \in [0, +\infty[, \\ v''(t) = h(t, u(t), v(t), u'(t), v'(t)), \end{cases}$$

with boundary conditions

$$u(0) = A_1, \ v(0) = A_2, \ u'(+\infty) = B_1, \ v'(+\infty) = B_2,$$

with $A_1, A_2, B_1, B_2 \in \mathbb{R}$, has a solution expressed by

$$u(t) = A_1 + B_1 t + \int_0^{+\infty} G(t, s) f(s, u(s), v(s), u'(s), v'(s)) ds$$

$$v(t) = A_2 + B_2 t + \int_0^{+\infty} G(t, s) h(s, u(s), v(s), u'(s), v'(s)) ds,$$

where

$$G(t, s) = \begin{cases} -t, & 0 \leq t \leq s \leq +\infty \\ \\ -s, & 0 \leq s \leq t \leq +\infty. \end{cases} \tag{3.10}$$

The next lemma gives a convenient criterion to guarantee the compacity on unbounded domains, and can be easily obtained from [6], Theorem 4.3.1, or [196], Theorem 2.3.

Theorem 3.1. *A set $M \subset X$ is relatively compact if the following conditions hold:*

(i) *both $\{t \to x(t) : x \in M\}$ and $\{t \to x'(t) : x \in M\}$ are uniformly bounded;*

(ii) *both $\{t \to x(t) : x \in M\}$ and $\{t \to x'(t) : x \in M\}$ are equicontinuous on any compact interval of $\mathbb{R}$;*

(iii) *both $\{t \to x(t) : x \subset M\}$ and $\{t \to x'(t) : x \in M\}$ are equiconvergent at $\pm\infty$, that is, for any given $\epsilon > 0$, there exists $t_\epsilon > 0$ such that*

$$\left| \frac{x(t)}{1+t} - \lim_{t \to +\infty} \frac{x(t)}{1+t} \right| < \epsilon, \left| x'(t) - \lim_{t \to +\infty} x'(t) \right| < \epsilon, \forall t > t_\epsilon, x \in M.$$

The existence tool will be given by Schauder's fixed point theorem (Theorem 5.2).

3.2 Existence result

In this section we prove the existence of solution for the problem (3.1)-(3.2).

Theorem 3.2. *Let $f, h : [0, +\infty[\times \mathbb{R}^4 \to \mathbb{R}$ be L^1-Carathéodory functions and assume that there is $R > 0$ such that*

$$R > \max \left\{ \max\{K_1, K_2\} + \int_0^{+\infty} K_3(s)\phi_R(s)ds, \right.$$

$$\left. \max\{|B_1|, |B_2|\} + \int_0^{+\infty} \phi_R(s)ds \right\}, \qquad (3.11)$$

where

$$K_i := \sup_{t \in [0, +\infty[} \left(\frac{|A_i| + |B_i t|}{1 + t} \right), \quad i = 1, 2, \quad K_3(s) := \sup_{t \in [0, +\infty[} \frac{|G(t, s)|}{1 + t}. \qquad (3.12)$$

Then there is at least a pair $(u, v) \in \left(C^2 [0, +\infty[\right)^2 \cap E$, solution of (3.1)-(3.2).

Proof. Define the operators $T_1 : E \to X$, $T_2 : E \to X$ and $T : E \to E$ given by

$$T(u, v) = (T_1(u, v), T_2(u, v)), \qquad (3.13)$$

with

$$(T_1(u, v))(t) = A_1 + B_1 t + \int_0^{+\infty} G(t, s)f(s, u(s), v(s), u'(s), v'(s))ds,$$

$$(T_2(u, v))(t) = A_2 + B_2 t + \int_0^{+\infty} G(t, s)h(s, u(s), v(s), u'(s), v'(s))ds,$$

where $G(t, s)$ is defined in (3.10).

The proof will follow several steps for clearness, only for operator $T_1(u, v)$. The technique for operator $T_2(u, v)$ is similar.

Step 1: T *is well defined and continuous.*

Let be $(u, v) \in E$. Therefore, it must be proved that $T(u, v) \in E$, that is $T_1(u, v) \in X$ and $T_2(u, v) \in X$.

As $(u, v) \in E$, then there exists some $r > 0$ such that $\|(u, v)\|_E \leq r$.

By the Lebesgue Dominated Theorem and Lemma 3.2,

$$
\begin{aligned}
\lim_{t \to +\infty} \frac{T_1\left(u,v\right)(t)}{1+t} &= \lim_{t \to +\infty} \frac{A_1 + B_1 t}{1+t} \\
&\quad + \int_0^{+\infty} \lim_{t \to +\infty} \frac{G(t,s)}{1+t} f(s, u(s), v(s), u'(s), v'(s)) ds \\
&\leq B_1 + \int_0^{+\infty} |f(s, u(s), v(s), u'(s), v'(s))| ds \\
&\leq B_1 + \int_0^{+\infty} \phi_r(s) ds < +\infty,
\end{aligned}
$$

and

$$
\begin{aligned}
\lim_{t \to +\infty} T_1\left(u,v\right)'(t) &= B_1 - \lim_{t \to +\infty} \int_t^{+\infty} f(s, u(s), v(s), u'(s), v'(s)) ds \\
&\leq B_1 + \lim_{t \to +\infty} \int_t^{+\infty} |f(s, u(s), v(s), u'(s), v'(s))| ds \\
&\leq B_1 + \int_0^{+\infty} |f(s, u(s), v(s), u'(s), v'(s))| ds \\
&\leq B_2 + \int_0^{+\infty} \phi_r(s) ds < +\infty.
\end{aligned}
$$

Therefore $T_1 \in X$. Analogously, $T_2 \in X$. So, T is well defined in E and, as f, h are L^1-Carathéodory functions, T is continuous.

Step 2: *TD is uniformly bounded, for D a bounded set in E.*

Let D be a bounded subset on E. Thus, there is $\rho_1 > 0$ such that

$$
\|(u,v)\|_E = \max\left\{\|u\|_X, \|v\|_X\right\} = \max\left\{\|u\|_0, \|u'\|_1, \|v\|_0, \|v'\|_1\right\} < \rho_1. \tag{3.14}
$$

As $0 \leq K_3(s) \leq 1$, $\forall s \in [0, +\infty[$ and f is a L^1-Carathéodory function, then

$$
\begin{aligned}
\|T_1\left(u,v\right)\|_0 &= \sup_{t \in [0,+\infty[} \frac{|T\left(u,v\right)(t)|}{1+t} \\
&\leq \sup_{t \in [0,+\infty[} \left(\frac{|A_1| + |B_1 t|}{1+t} \right) \\
&\quad + \int_0^{+\infty} \sup_{t \in [0,+\infty[} \frac{|G(t,s)|}{1+t} |f(s, u(s), v(s), u'(s), v'(s))| ds \\
&\leq K_1 + \int_0^{+\infty} K_3(s) \phi_{\rho_1}(s) ds < +\infty, \forall\, (u,v) \in D,
\end{aligned}
$$

and

$$\| (T_1 (u, v))' \|_1 = \sup_{t \in [0, +\infty[} | (T_1 (u, v))' (t) |$$

$$\leq |B_1| + \int_t^{+\infty} |f(s, u(s), v(s), u'(s), v'(s))| ds$$

$$\leq |B_1| + \int_0^{+\infty} \phi_{\rho_1} (s) ds < +\infty, \forall (u, v) \in D.$$

Taking into account these arguments, T_2 verifies similar bounds and $\|T(u, v)\|_E < \rho_1$, that is TD is uniformly bounded.

Step 3: TD *is equicontinuous in* E. Let $t_1, t_2 \in [0, +\infty[$ and suppose, without loss of generality, that $t_1 \leq t_2$. So, by the continuity of $G(t, s)$,

$$\lim_{t_1 \to t_2} \left| \frac{T_1 (u, v) (t_1)}{1 + t_1} - \frac{T_1 (u, v) (t_2)}{1 + t_2} \right| \leq \lim_{t_1 \to t_2} \left| \left(\frac{A_1 + B_1 t_1}{1 + t_1} - \frac{A_1 + B_1 t_2}{1 + t_2} \right) \right|$$

$$+ \int_0^{+\infty} \lim_{t_1 \to t_2} \left| \frac{G(t_1, s)}{1 + t_1} - \frac{G(t_2, s)}{1 + t_2} \right| |f(s, u(s), v(s), u'(s), v'(s))| ds = 0,$$

and

$$\lim_{t_1 \to t_2} \left| (T_1 (u, v) (t_1))' - (T_1 (u, v) (t_2))' \right|$$

$$= \lim_{t_1 \to t_2} \left| - \int_{t_1}^{+\infty} f(s, u(s), v(s), u'(s), v'(s)) ds \right.$$

$$\left. + \int_{t_2}^{+\infty} f(s, u(s), v(s), u'(s), v'(s)) ds \right|$$

$$= \lim_{t_1 \to t_2} \left| - \int_{t_1}^{t_2} f(s, u(s), v(s), u'(s), v'(s)) ds \right| \leq \lim_{t_1 \to t_2} \int_{t_1}^{t_2} \phi_{\rho_1} (s) ds = 0.$$

Therefore, $T_1 D$ is equicontinuous on E. In the same way it can be proved that $T_2 D$ is equicontinuous on E, too. Thus, TD is equicontinuous on E.

Step 4: TD *is equiconvergent at infinity.* For the operator T_1, we have

$$\left| \frac{T_1 (u, v) (t)}{1 + t} - \lim_{t \to +\infty} \frac{T_1 (u, v) (t)}{1 + t} \right| \leq \left| \frac{A_1 + B_1 t}{1 + t} - B_1 \right|$$

$$+ \int_0^{+\infty} \left| \frac{G(t, s)}{1 + t} - \lim_{t \to +\infty} \frac{G(t, s)}{1 + t} \right| |f(s, u(s), v(s), u'(s), v'(s)) ds| \to 0,$$

as $t \to +\infty$.

Similarly,

$$\left| (T_1(u,v)(t))' - \lim_{t \to +\infty} (T_1(u,v)(t))' \right|$$

$$= \left| -\int_t^{+\infty} f(s,u(s),v(s),u'(s),v'(s))ds \right|$$

$$\leq \int_t^{+\infty} |f(s,u(s),v(s),u'(s),v'(s))|ds$$

$\to 0$, as $t \to +\infty$.

$T_1 D$ is equiconvergent at $+\infty$ and, following a similar technique, we can prove that $T_2 D$ is equiconvergent at $+\infty$, too. So, TD is equiconvergent at $+\infty$. By Theorem 3.1, TD is relatively compact and, consequently, T is completely continuous.

Step 5: $T\Omega \subset \Omega$ for some $\Omega \subset E$ *a closed and bounded set.*

Consider

$$\Omega := \{(u,v) \in E : \|(u,v)\|_E \leq \rho_2\},$$

with $\rho_2 > 0$ such that

$$\rho_2 := \max \left\{ \begin{array}{c} \rho_1, \ K_1 + \int_0^{+\infty} K_3(s)\phi_{\rho_2}(s)ds, \ K_2 + \int_0^{+\infty} K_3(s)\phi_{\rho_2}(s)ds, \\[2mm] |B_1| + \int_0^{+\infty} \phi_{\rho_2}(s)ds, \ |B_2| + \int_0^{+\infty} \phi_{\rho_2}(s)ds \end{array} \right\},$$

with ρ_1 given by (3.14), according to Step 2 and K_1, K_2 and $K_3(s)$ given by (3.12), we have

$$\|T(u,v)\|_E = \|(T_1(u,v), T_2(u,v))\|_E$$
$$= \max \{\|T_1(u,v)\|_X, \|T_2(u,v)\|_X\}$$
$$= \max \{\|T_1(u,v)\|_0, \|(T_1(u,v))'\|_1,$$
$$\|T_2(u,v)\|_0, \|(T_2(u,v))'\|_1\} \leq \rho_2.$$

So, $T\Omega \subset \Omega$, and by Schauder's Theorem (Theorem 5.2), the operator $T(u,v) = (T_1(u,v), T_2(u,v))$, has a fixed point (u,v). By standard techniques it can be shown that this fixed point is a solution of problem (3.1)-(3.2). $\qquad\square$

3.3   Application to a predator-prey model

Mathematical models have been considered to study the dynamics of some species and organisms living in environments where there are strong unidirectional flows, such as a river or a stream. An important goal is to understand how populations and ecosystems can survive in such media.

Some examples: predator-prey dynamics can be studied via reaction-diffusion-advection systems (for details see [271]); the effect of network structures and discrete diffusion rates on predator-prey-subsidy dynamics of Stepping-Stone models [236]; a predator-prey model with switching between a traditional model (the free system) and a model with a nonlinear harvesting regime for the predator population (the harvesting system) [255]; the spatially heterogeneous subsidy distributions, focusing on the potential they have for destabilizing coexistence equilibria between predator-prey interactions, leading to ecological cascade effects [38].

In this section we consider an application to a stationary predator-prey model in a very long domain, given by the second-order nonlinear coupled system on the half-line, composed by the differential equations

$$
\begin{cases}
u''(x) = \left(\frac{b_1}{\gamma_1} u'(x) - \frac{1}{\gamma_1} u(x) \left[r_1 - b_{11} u(x) - b_{12} v(x) \right] \right) \frac{1}{1+x^4} \\[2mm]
v''(x) = \left(\frac{b_2}{\gamma_2} v'(x) - \frac{1}{\gamma_2} v(x) \left[r_2 - b_{21} v(x) - b_{22} u(x) \right] \right) \frac{1}{1+x^4},
\end{cases}
\tag{3.15}
$$

where:

- $u(x)$ and $v(x)$ denote the densities of the prey and the predator populations, respectively, at position $x \in [0, +\infty[$;
- $\gamma_1, \gamma_2 > 0$, are diffusion constants of the prey and the predator, respectively;
- $b_1, b_2 \in \mathbb{R}$, are the advection rates of the prey and the predator, respectively;
- $r_1 > 0$, $r_2 \in \mathbb{R}$, are the growth rates of the prey and the predator, respectively;
- $b_{11}, b_{21} > 0$, are the density-dependent constants of the prey and the predator, respectively;
- $b_{12}, b_{22} > 0$, are the predation and conversion rates, respectively;

together the boundary conditions

$$
\begin{cases}
u(0) = A_1, \ v(0) = A_2, \\
u'(+\infty) = B_1, \ v'(+\infty) = B_2,
\end{cases}
\tag{3.16}
$$

with $A_1, A_2 > 0$, $B_1, B_2 \in \mathbb{R}$.

Remark that:

1. B_i are real numbers, for $i = 1, 2$. Therefore, this model may consider cases of growth, decay or constant populations of preys and predators;
2. The derivatives at infinity provide the variation of preys and predators with increasing distance, this is, $x \to +\infty$. The possibility of having infinite space is important, especially when the population size of prey becomes small or when populations lose spatial contact;
3. $u(x) = 0$, meaning the absence of prey, will lead to the extinction of the predator population due to lack of food;
4. $v(x) = 0$, means the absence of the predator. The population of prey will increase, without any type of obstacle, till an eventually resource deflection.

The system (3.15)-(3.16) is a particular case of the problem (3.1)-(3.2), with

$$f(x, m, y, z, w) = \left(\frac{b_1}{\gamma_1} z - \frac{1}{\gamma_1} m \left[r_1 - b_{11}m - b_{12}y \right] \right) \frac{1}{1 + x^4},$$

and

$$h(x, m, y, z, w) = \left(\frac{b_2}{\gamma_2} w - \frac{1}{\gamma_2} y \left[r_2 - b_{21}y - b_{22}m \right] \right) \frac{1}{1 + x^4}.$$

In fact, these functions are L^1-Carathéodory functions, with

$$f(x, m, y, z, w) \leq \left(\frac{b_1}{\gamma_1} \rho + \frac{1}{\gamma_1} \rho^2 (1 + x)^2 (b_{11} + b_{12}) + \rho(1 + x) r_1 \right) \frac{1}{1 + x^4}$$
$$:= \phi_\rho(x),$$

and

$$h(x, m, y, z, w) \leq \left(\frac{b_2}{\gamma_2} \rho + \frac{1}{\gamma_2} \rho^2 (1 + x)^2 (b_{21} + b_{22}) + \rho(1 + x) r_2 \right) \frac{1}{1 + x^4}$$
$$:= \varphi_\rho(x),$$

for some $\rho > 0$, such that

$$\sup_{x \in [0, +\infty[} \left\{ \frac{|m|}{1 + x}, \frac{|y|}{1 + x}, |z|, |w| \right\} < \rho.$$

So, by Theorem 3.2, there is at least a pair $(u, v) \in \left(C^2 [0, +\infty[\right)^2 \cap E$, solution of (3.15)-(3.16).

3.4 Existence and localization result

In this section we shall prove the existence and localization of a solution for problem (3.1)-(3.2).

Let $A_1, A_2, B_1, B_2 \in \mathbb{R}$. Lower and upper functions are defined in the following way:

Definition 3.3. A pair of functions $(\alpha_1, \alpha_2) \in \left(C^2\,[0, +\infty[\right)^2 \cap E$ is a lower solution of problem (3.1), (3.2) if

$$\alpha_1''(t) \leq f(t, \alpha_1(t), \alpha_2(t), \alpha_1'(t), w), \forall w \in \mathbb{R}$$
$$\alpha_2''(t) \leq h(t, \alpha_1(t), \alpha_2(t), z, \alpha_2'(t)), \forall z \in \mathbb{R}$$
$$\alpha_1(0) \leq A_1$$
$$\alpha_2(0) \leq A_2$$
$$\alpha_1'(+\infty) < B_1$$
$$\alpha_2'(+\infty) < B_2.$$

A pair of functions $(\beta_1, \beta_2) \in \left(C^2\,[0, +\infty[\right)^2 \cap E$ is an upper solution of problem (3.1), (3.2) if it verifies the reverse inequalities.

Theorem 3.3. *Let $f, h : [0, +\infty[\times \mathbb{R}^4 \to \mathbb{R}$ be L^1-Carathéodory functions, the assumptions of Theorem 3.2 and the Nagumo-type conditions given by Definition 3.2, hold.*

If there are coupled lower and upper solutions of (3.1)-(3.2), (α_1, α_2) and (β_1, β_2), respectively, such that

$$\alpha_1(t) \leq \beta_1(t), \ \alpha_2(t) \leq \beta_2, \ \forall t \in [0, +\infty[,$$

then there is at least a pair $(u(t), v(t)) \in \left(C^2\,[0, +\infty[\right)^2 \cap E$ solution of (3.1)-(3.2), such that

$$\alpha_1(t) \leq u(t) \leq \beta_1(t), \ \alpha_2(t) \leq v(t) \leq \beta_2(t), \ \forall t\,[0, +\infty[, \tag{3.17}$$

and

$$\|u'\| \leq N_1 \ and \ \|v'\| \leq N_2,$$

with N_1 and N_2 given by Lemma 3.1.

Proof. Consider the operator T given by (3.13). By Theorem 3.2, there is a fixed point of T, (u, v), which is a solution of problem (3.1)-(3.2).

To prove the localization part given by (3.17), consider the auxiliary functions $\delta_i : [0, +\infty[\times \mathbb{R} \to \mathbb{R}$, $i = 1, 2$, defined as

$$\delta_i(t, w) = \begin{cases} \beta_i(t), & w > \beta_i(t) \\ w, & \alpha_i(t) \leq w \leq \beta_i(t) \\ \alpha_i(t), & w < \alpha_i(t), \end{cases}$$

and the truncated and perturbed coupled system

$$\begin{cases} u''(t) = f(t, \delta_1(t, u(t)), \delta_2(t, v(t)), u'(t), v'(t)) + \frac{1}{1+t} \frac{u(t) - \delta_1(t, u(t))}{|u(t) - \delta_1(t, u(t))|}, \\ v''(t) = h(t, \delta_1(t, u(t)), \delta_2(t, v(t)), u'(t), v'(t)) + \frac{1}{1+t} \frac{v(t) - \delta_2(t, v(t))}{|v(t) - \delta_2(t, v(t))|}. \end{cases}$$

Suppose, by contradiction, that there is $t \in [0, +\infty[$, such that $\alpha_1(t) > u(t)$ and define

$$\inf_{t \in [0, +\infty[} (u(t) - \alpha_1(t)) := u(t_0) - \alpha_1(t_0) < 0.$$

Then, $t_0 \neq 0$ and $t_0 \neq +\infty$, as, by Definition 3.3 and (3.2),

$$u(0) - \alpha_1(0) = A_1 - \alpha_1(0) \geq 0,$$

and

$$u'(+\infty) - \alpha_1'(+\infty) = B_1 - \alpha_1'(+\infty) > 0.$$

Therefore, $t_0 \in]0, +\infty[$, $u'(t_0) = \alpha_1'(t_0)$, $u''(t_0) - \alpha_1''(t_0) \geq 0$. So, we deduce the follow contradiction

$$0 \leq u''(t_0) - \alpha_1''(t_0)$$
$$= f(t_0, \delta_1(t_0, u(t_0)), \delta_2(t_0, v(t_0)), u'(t_0), v'(t_0))$$
$$+ \frac{1}{1 + t_0} \frac{u(t_0) - \delta_1(t, u(t_0))}{|u(t_0) - \delta_1(t, u(t_0))| + 1} - \alpha_1''(t_0)$$
$$\leq f(t_0, \alpha_1(t_0), \alpha_2(t_0), \alpha_1'(t_0), v'(t_0)) + \frac{1}{1 + t_0} \frac{u(t_0) - \alpha_1(t_0)}{|u(t_0) - \alpha_1(t_0)| + 1} - \alpha_1''(t_0)$$
$$\leq \frac{1}{1 + t_0} \frac{u(t_0) - \alpha_1(t_0)}{|u(t_0) - \alpha_1(t_0)| + 1} < 0.$$

So, $\alpha_1(t) \leq u(t)$, $\forall t \in [0, +\infty[$, and the remaining inequalities $u(t) \leq \beta_1(t)$, $\forall t \in [0, +\infty[$, can be proved by the same technique.

Applying the method above, it may be shown that $\alpha_2(t) \leq v(t) \leq \beta_2(t)$, $\forall t \in [0, +\infty[$. $\qquad\square$

3.5 Example

Consider the second-order coupled system

$$\begin{cases} u''(t) = \dfrac{\pi u'(t)\left[u(t) - 3\sqrt[3]{v(t)} + \arctan\left(v'(t)\right)\right]}{t^4 + 1}, & t \in [0, +\infty[, \\ v''(t) = \dfrac{v'(t)\left[2e^{-|u'(t)|} - (u(t))^3 - \sqrt[3]{v(t)}\right]}{1 + t^6}, & t \in [0, +\infty[\end{cases} \tag{3.18}$$

and the boundary conditions

$$u(0) = v(0) = 0, \quad u'(+\infty) = 1, \quad v'(+\infty) = 1. \tag{3.19}$$

In fact, this problem is a particular case of system (3.1)-(3.2), with

$$f(t, x, y, z, w) = \frac{\pi z\left[x - 3\sqrt[3]{y} + \arctan\left(w\right)\right]}{t^4 + 1}$$

$$\leq \frac{\pi \rho\left[(t+1)\rho + 3\sqrt[3]{\rho(1+t)} + \frac{\pi}{2}\right]}{t^4 + 1} := \phi_\rho(t),$$

and

$$h(t, x, y, z, w) = \frac{w\left[2e^{-|z|} + x^3 + \sqrt[3]{y}\right]}{1 + t^6}$$

$$\leq \frac{\rho\left[2 + (1+t)^3\rho^3 + \sqrt[3]{(1+t)\rho}\right]}{1 + t^6} := \phi_\rho(t),$$

for some $\rho > 0$, such that

$$\sup_{t \in [0, +\infty[} \left\{ \frac{|x(t)|}{1+t}, \frac{|y(t)|}{1+t}, |z(t)|, |w(t)| \right\} < \rho.$$

Therefore, $f, h : [0, +\infty[\times \mathbb{R}^4 \to \mathbb{R}$ are L^1-Carathéodory functions.

The functions given by

$$(\alpha_1(t), \alpha_2(t)) = (-2 - t, -1 + t) \text{ and } (\beta_1(t), \beta_2(t)) = (1 + t, 2 + t)$$

are, respectively, lower and upper solutions of problem (3.18)-(3.19), with $\alpha_1(t) = -2 - t \leq \beta_1(t) = 1 + t$ and $\alpha_2(t) = -1 + t \leq \beta_2(t) = 2 + t$, $\forall t \in [0, +\infty[$.

Furthermore, the functions

$$f(t,x,y,z,w) \le \frac{\pi|z|\left[(t+1) + 3\sqrt[3]{2+t} + \frac{\pi}{2}\right]}{t^4 + 1}$$

and

$$h(t,x,y,z,w) = \frac{|w|\left[2e^{-|z|} + (1+t)^3 + \sqrt[3]{(2+t)}\right]}{1 + t^6}$$

$$\le \frac{|w|\left[2 + (1+t)^3 + \sqrt[3]{2+t}\right]}{1 + t^6},$$

satisfy a Nagumo condition relative to the set

$$S = \{(t,x,y,z,w) \in [0, +\infty[\, \times \mathbb{R}^4 : -2-t \le x \le 1+t, \ -1+t \le y \le 2+t\},$$

with

$$\phi(t) := \frac{(t+1) + 3\sqrt[3]{2+t} + \frac{\pi}{2}}{t^4 + 1}, \quad l_1(|z|) = \pi|z|,$$

$$\varphi(t) := \frac{2 + (1+t)^3 + \sqrt[3]{2+t}}{1 + t^6}, \quad l_2(|w|) = |w|.$$

We can also see that,

$$\int_0^{+\infty} \phi(s)ds = \int_0^{+\infty} \frac{(s+1) + 3\sqrt[3]{2+s} + \frac{\pi}{2}}{s^4 + 1} ds \simeq 8.26013 < +\infty,$$

$$\int_0^{+\infty} \varphi(s)ds = \int_0^{+\infty} \frac{2 + (1+s)^3 + \sqrt[3]{2+s}}{1 + s^6} ds \simeq 8.56284 < +\infty,$$

$$\int_0^{+\infty} \frac{s}{l_1(s)}ds = \int_0^{+\infty} \frac{s}{\pi s}ds = +\infty; \quad \int_0^{+\infty} \frac{s}{l_2(s)}ds = \int_0^{+\infty} ds = +\infty,$$

and, for $1 < \varepsilon \le 3$,

$$\sup_{t\in[0,+\infty[} \phi(t)(1+t)^\varepsilon = \sup_{t\in[0,+\infty[} \frac{(t+1) + 3\sqrt[3]{2+t} + \frac{\pi}{2}}{t^4 + 1}(1+t)^\varepsilon$$

$$\simeq 31.59018 \le R_1,$$

and

$$\sup_{t\in[0,+\infty[} \varphi(t)(1+t)^\varepsilon = \sup_{t\in[0,+\infty[} \frac{2 + (1+t)^3 + \sqrt[3]{2+t}}{1 + t^6}(1+t)^\varepsilon$$

$$\simeq 45.76900 \le R_2.$$

Thereby, by Theorem 3.2, there is at least a pair $(u(t), v(t))$ $\in \left(C^2([0, +\infty[), \mathbb{R}\right)^2$ solution of (3.18)-(3.19) and, moreover,

$$-2 - t \leq u(t) \leq 1 + t, \quad -1 + t \leq v(t) \leq 2 + t, \; \forall t \in [0, +\infty[,$$

and there are positive N_1, N_2, such that $\|u'\|_1 \leq N_1$ and $\|v'\|_1 \leq N_2$.

Chapter 4

Homoclinic solutions for second-order coupled systems

In this chapter, we consider the second-order coupled system on the real line

$$
\begin{cases}
u''(t) - k_1^2 u(t) = f(t, u(t), v(t), u'(t), v'(t)), \ t \in \mathbb{R}, \\
v''(t) - k_2^2 v(t) = h(t, u(t), v(t), u'(t), v'(t)), \ t \in \mathbb{R},
\end{cases}
\tag{4.1}
$$

with $f, h : \mathbb{R}^5 \to \mathbb{R}$ L^1-Carathéodory functions, $k_1, k_2 > 0$, and the boundary conditions

$$
\begin{cases}
u(\pm\infty) = 0, \ v(\pm\infty) = 0, \\
u'(\pm\infty) = 0, \ v'(\pm\infty) = 0,
\end{cases}
\tag{4.2}
$$

where,

$$
u(\pm\infty) := \lim_{t \to \pm\infty} u(t), \quad v(\pm\infty) := \lim_{t \to \pm\infty} v(t),
$$

$$
u'(\pm\infty) := \lim_{t \to \pm\infty} u'(t), \quad v'(\pm\infty) := \lim_{t \to \pm\infty} v'(t).
$$

Moreover, we present sufficient conditions for the existence of homoclinic solutions, based on [198].

An homoclinic orbit is a flow trajectory of a dynamical system that joins a saddle equilibrium point into itself, that is, the homoclinic trajectory converges at an equilibrium point as $t \to \pm\infty$. So, by an homoclinic solution of system (4.1), we mean a nontrivial solution of (4.1) verifying (4.2).

The importance of homoclinics is described in [8, 86, 179, 182, 219], with applications involving dynamic systems, namely, dynamics of coupled cell networks, bifurcation theory, chaos and homoclinic orbits near resonances in reversible systems.

In [61, 62], Champneys and Lord study homoclinic solutions of periodic orbits in a reduced water-wave problem, in reversible systems and their applications in mechanics, fluids and optics. Algaba *et al.*, in [12], deal with an application to Rössler system. In [90], Fečkan presents homoclinic orbits with Melnikov Mappings, difference equation and planar perturbations.

Results about the existence of homoclinic multiple solutions can be seen, for example: in [257], where Wilczak shows the existence of infinitely many homoclinic and heteroclinic connections between two odd periodic solutions of the Michelson system $x''' + x' + 0.5x^2 = 1$; in [246], Sun and Wu present some new results of homoclinic solutions and the existence of two different homoclinic solutions for second-order Hamiltonian systems; in [245], by establishing a compactness lemma and using variational methods, the author proves the existence of multiple homoclinic solutions for a class of fourth-order differential equations with a perturbation; in [150], based on the fountain theorem in combination with variational technique, Kong obtained a sufficient condition for the existence of infinitely many homoclinic solutions of higher order difference equation with p-Laplacian and containing both advances and delays.

In [63], it is proved the existence and multiplicity for infinitely many homoclinic orbits from 0 of the second-order Hamiltonian system

$$\ddot{u} - a(t)|u(t)|^{p-2}u(t) + \nabla W(t, u(t)) = 0,$$

where $p \geq 2$, $t \in \mathbb{R}$, $u \in \mathbb{R}^N$, $a : \mathbb{R} \to \mathbb{R}$ and $W : \mathbb{R} \times \mathbb{R}^N \to \mathbb{R}$.

In [178], Liu, Guo and Zhang, study the existence and also the multiplicity of homoclinic orbits for the following second-order Hamiltonian systems

$$\ddot{u} - L(t)u + \nabla W(t, u) = 0,$$

where $L \in C(\mathbb{R}, \mathbb{R}^{N^2})$ is a symmetric and positive-definite matrix for all $t \in \mathbb{R}$, $W \in C^1(\mathbb{R} \times \mathbb{R}^N, \mathbb{R})$ and $\nabla W(t, u)$ is the gradient of W with respect to u.

Motivated by the works above and based in the arguments used in [196], we apply the fixed point theory, the lower and upper solutions method combined with adequate growth assumptions on the nonlinearities, to obtain sufficient conditions for the existence of homoclinic solutions of the coupled system (4.1). Moreover, our technique provides a strip where both homoclinics lie.

On the other hand, there are many phenomena that are not modeled by differential equations, but by systems. The family of second-order nonlinear coupled systems of type (4.1) can model various phenomena. For example, in [230], the authors analyze the peak solutions for a reaction-diffusion model; to study the binary fluid convection model (see [228]); example from the geophysical morphodynamics (see [148]).

We emphasize that the vast area of applications for this family of second-order nonlinear coupled systems certainly warrants a study of the coupled system (4.1).

Note that it is the first time where homoclinic solutions for second-order coupled differential systems is considered with differential equations having full nonlinearities on both unknown functions. We point out that although the arguments are similar to [196], for coupled systems the technique is more delicate as it requires stronger definitions of lower and upper solutions due to the dependence on both unknown functions and their first derivative (see Definition 4.1).

4.1 Preliminary results

Consider the following spaces

$$X_1 := \left\{ x \in C^1(\mathbb{R}) : \lim_{|t| \to \infty} x^{(i)}(t) \in \mathbb{R}, \ i = 0, 1 \right\},$$

and $X := X_1 \times X_1$, equipped with the norm

$$\|x\|_{X_1} = \max \left\{ \|x\|_\infty, \|x'\|_\infty \right\},$$

where

$$\|w\|_\infty := \sup_{t \in \mathbb{R}} |w(t)|,$$

and

$$\|(u, v)\|_X = \max \left\{ \|u\|_{X_1}, \|v\|_{X_1} \right\}.$$

It can be proved that, $(X_1, \|\cdot\|_{X_1})$ and $(X, \|\cdot\|_X)$ are Banach spaces.

For the reader's convenience we precise that L^1-Carathéodory functions are given by Definition 3.1, considering now a function $g : \mathbb{R}^5 \to \mathbb{R}$ and replacing condition *(iii)* by:

For each $\rho > 0$, there exists a positive function $\phi_\rho \in L^1(\mathbb{R})$ such that, whenever $x, y, z, w \in [-\rho, \rho]$, then

$$|g(t, x, y, z, w)| \le \phi_\rho(t), \text{ a.e. } t \in \mathbb{R}. \tag{4.3}$$

The next lemma guarantees the existence of the solution of problem (4.1), (4.2) in $L^1(\mathbb{R})$.

Lemma 4.1. *Let $f^*, h^* \in L^1(\mathbb{R})$. Then the system*

$$\begin{cases} u''(t) - k_1^2 u(t) = f^*(t), & \text{a.e. } t \in \mathbb{R} \\ v''(t) - k_2^2 v(t) = h^*(t), \end{cases} \tag{4.4}$$

with conditions (4.2), has a unique solution expressed by

$$u(t) = \int_{-\infty}^{+\infty} G_1(t, s) f^*(s) ds,$$

$$v(t) = \int_{-\infty}^{+\infty} G_2(t, s) h^*(s) ds,$$

where

$$G_1(t, s) = \begin{cases} -\frac{1}{2k_1} e^{k_1(s-t)}, & -\infty \le s \le t \le +\infty, \\ -\frac{1}{2k_1} e^{k_1(t-s)}, & -\infty \le t \le s \le +\infty, \end{cases} \tag{4.5}$$

and

$$G_2(t, s) = \begin{cases} -\frac{1}{2k_2} e^{k_2(s-t)}, & -\infty \le s \le t \le +\infty, \\ -\frac{1}{2k_2} e^{k_2(t-s)}, & -\infty \le t \le s \le +\infty. \end{cases} \tag{4.6}$$

Proof. The solution of the homogeneous differential system

$$\begin{cases} u''(t) - k_1^2 u(t) = 0, \\ v''(t) - k_2^2 v(t) = 0, \end{cases}$$

is

$$\left(u(t), \, v(t) \right) = \left(c_1 e^{k_1^2 t} + c_2 e^{-k_1^2 t}, \; c_3 e^{k_2^2 t} + c_4 e^{-k_2^2 t} \right),$$

with $c_1, c_2, c_3, c_4 \in \mathbb{R}$.

By the method of variation of parameters, the general solution of (4.4), (4.2) is given by

$$u(t) = -\int_{-\infty}^{t} \frac{e^{k_1(s-t)}}{2k_1} f^*(s) ds - \int_{t}^{+\infty} \frac{e^{k_1(t-s)}}{2k_1} f^*(s) ds,$$

$$v(t) = -\int_{-\infty}^{t} \frac{e^{k_2(s-t)}}{2k_2} h^*(s) ds - \int_{t}^{+\infty} \frac{e^{k_2(t-s)}}{2k_2} h^*(s) ds,$$

and, therefore,

$$u(t) = \int_{-\infty}^{+\infty} G_1(t,s) f^*(s) ds,$$

$$v(t) = \int_{-\infty}^{+\infty} G_2(t,s) h^*(s) ds,$$

with G_1 and G_2 given by (4.5) and (4.6), respectively. $\qquad \square$

To guarantee a convenient criterion for compacity, we consider Theorem 3.1, with condition *(iii)* replaced by:

$$|f(t) - f(\pm\infty)| < \epsilon, \ |f'(t) - f'(\pm\infty)| < \epsilon, \forall |t| > t_\epsilon, f \in M.$$

Strict lower and upper solutions of problem (4.1), (4.2) are defined in the following way:

Definition 4.1. A pair of functions $(\alpha_1, \alpha_2) \in X$ is a strict lower solution of problem (4.1), (4.2) if

$$\begin{aligned}
\alpha_1''(t) - k_1^2 \alpha_1(t) &> f(t, \alpha_1(t), \alpha_2(t), \alpha_1'(t), w), \ t, w \in \mathbb{R}, \\
\alpha_2''(t) - k_2^2 \alpha_2(t) &> h(t, \alpha_1(t), \alpha_2(t), z, \alpha_2'(t)), \ t, z \in \mathbb{R}, \\
\alpha_1(\pm\infty) &\leq 0, \\
\alpha_2(\pm\infty) &\leq 0.
\end{aligned} \qquad (4.7)$$

A pair of functions $(\beta_1, \beta_2) \in X$ is a strict upper solution of problem (4.1), (4.2) if it verifies the reverse inequalities.

The existence tool will be given by Theorem 5.2.

Moreover, along this work we denote $(a, b) \leq (c, d)$ meaning that $a \leq c$ and $b \leq d$, for $a, b, c, d \in \mathbb{R}$.

4.2 Existence and localization of homoclinics

In this section we prove the existence and localization for a solution of the problem (4.1)-(4.2). Moreover, by homoclinic solutions, we mean a non trivial solution of (4.1)-(4.2).

Theorem 4.1. *Let $f, h : \mathbb{R}^5 \to \mathbb{R}$ be L^1-Carathéodory functions with $f(t, 0, 0, 0, 0) \neq 0$, $h(t, 0, 0, 0, 0) \neq 0$, for some $t \in \mathbb{R}$, and there is $R > 0$ such that*

$$\max \left\{ \int_{-\infty}^{+\infty} M_1(s) \psi_R(s) ds < +\infty, \int_{-\infty}^{+\infty} M_2(s) \psi_R^*(s) ds < +\infty, \right\} < R$$

where

$$M_i(s) := \max \left\{ \sup_{t \in \mathbb{R}} |G_i(t,s)|, \sup_{t \in \mathbb{R}} \left| \frac{\partial G_i(t,s)}{\partial t} \right|, \ i = 1,2, \right\},$$

and, for $\rho > 0$, $x, y, z, w \in [-\rho, \rho]$,

$$|f(t,x,y,z,w)| \le \psi_\rho(t), \ \text{a.e.} \ t \in \mathbb{R}, \tag{4.8}$$

$$|h(t,x,y,z,w)| \le \psi_\rho^*(t), \ \text{a.e.} \ t \in \mathbb{R}. \tag{4.9}$$

Moreover if (α_1, α_2), $(\beta_1, \beta_2) \in X$ are lower and upper solutions of problem (4.1)-(4.2), respectively, such that

$$(\alpha_1(t), \alpha_2(t)) \le (\beta_1(t), \beta_2(t)), \ \forall t \in \mathbb{R}, \tag{4.10}$$

and, if $f(t,x,y,z,w)$ is nonincreasing on y and monotone (nonincreasing or nondecreasing) on z, for fixed $(t,x,w) \in \mathbb{R}^3$, and $h(t,x,y,z,w)$ is nonincreasing in x and monotone (nonincreasing or nondecreasing) on w, for fixed $(t,y,z) \in \mathbb{R}^3$, then the problem (4.1)-(4.2) has a homoclinic solution $(u,v) \in X$ such that

$$\alpha_1(t) \le u(t) \le \beta_1(t), \quad \alpha_2(t) \le v(t) \le \beta_2(t), \ \forall t \in \mathbb{R}.$$

Proof. Consider the auxiliary functions $\delta_i : \mathbb{R}^2 \to \mathbb{R}$, for $i = 1, 2$, given by

$$\delta_i(t,w) = \begin{cases} \beta_i(t), & w > \beta_i(t) \\ w, & \alpha_i(t) \le w \le \beta_i(t) \\ \alpha_i(t), & w < \alpha_i(t), \end{cases} \tag{4.11}$$

and the truncated and perturbed coupled system

$$\begin{cases} u''(t) - k_1^2 u(t) = f(t, \delta_1(t, u(t)), \delta_2(t, v(t)), u'(t), v'(t)), \\ v''(t) - k_2^2 v(t) = h(t, \delta_1(t, u(t)), \delta_2(t, v(t)), u'(t), v'(t)). \end{cases} \tag{4.12}$$

Define the operators $T_1, T_2 : X \to X_1$, and $T : X \to X$ by

$$T(u,v) = (T_1(u,v), T_2(u,v)), \tag{4.13}$$

with

$$(T_1(u,v))(t) = \int_{-\infty}^{+\infty} G_1(t,s) f(s, \delta_1(s, u(s)), \delta_2(s, v(s)), u'(s), v'(s)) ds,$$

$$(T_2(u,v))(t) = \int_{-\infty}^{+\infty} G_2(t,s) h(s, \delta_1(s, u(s)), \delta_2(s, v(s)), u'(s), v'(s)) ds,$$

where $G_1(t,s)$, $G_2(t,s)$ are defined by (4.5) and (4.6), respectively.

By Lemma 4.1, the fixed points of T are solutions of (4.1)-(4.2) and from Theorem 5.2, we need to show that the operator T is compact, in order to prove that $T(u, v)$ has a fixed point.

For clearness, we divide this proof into several steps. The technique for operator $T_2(u, v)$ is similar.

(i) *T is well defined and continuous in X.*

Let $(u, v) \in X$. As f, h are L^1-Carathéodory functions, T is continuous and by the Lebesgue dominated theorem,

$$\lim_{t \to \pm\infty} T_1(u, v)(t)$$

$$= \lim_{t \to \pm\infty} \int_{-\infty}^{+\infty} G_1(t, s) f(s, \delta_1(s, u(s)), \delta_2(s, v(s)), u'(s), v'(s)) ds$$

$$= \lim_{t \to -\infty} \int_{-\infty}^{t} \frac{-1}{2k_1} e^{k_1(s-t)} f(s, \delta_1(s, u(s)), \delta_2(s, v(s)), u'(s), v'(s)) ds$$

$$+ \lim_{t \to +\infty} \int_{t}^{+\infty} \frac{-1}{2k_1} e^{k_1(t-s)} f(s, \delta_1(s, u(s)), \delta_2(s, v(s)), u'(s), v'(s)) ds = 0.$$

As

$$\frac{\partial G_1(t, s)}{\partial t} = \begin{cases} \frac{1}{2} e^{k_1(s-t)}, & -\infty \leq s < t \leq +\infty \\ -\frac{1}{2} e^{k_1(t-s)}, & -\infty \leq t < s \leq +\infty, \end{cases} \tag{4.14}$$

then

$$\lim_{t \to \pm\infty} (T_1(u, v)(t))'$$

$$= \lim_{t \to \pm\infty} \int_{-\infty}^{+\infty} \frac{\partial G_1(t, s)}{\partial t} f(s, \delta_1(s, u(s)), \delta_2(s, v(s)), u'(s), v'(s)) ds$$

$$= \lim_{t \to -\infty} \int_{-\infty}^{t} \frac{-1}{2} e^{k_1(s-t)} f(s, \delta_1(s, u(s)), \delta_2(s, v(s)), u'(s), v'(s)) ds$$

$$+ \lim_{t \to +\infty} \int_{t}^{+\infty} \frac{1}{2} e^{k_1(t-s)} f(s, \delta_1(s, u(s)), \delta_2(s, v(s)), u'(s), v'(s)) ds = 0.$$

Therefore, $T_1(u, v) \in X_1$. For $T_2(u, v) \in X_1$ the arguments are similar and, therefore, $T(u, v) \in X$.

(ii) *TB is uniformly bounded on $B \subseteq X$, for some bounded B.*

As f, h are L^1-Carathéodory functions, there exist ψ_ρ, $\psi_\rho^* \in L^1(\mathbb{R})$, with $\rho > 0$, such that,

$$\rho > \max\left\{\|\alpha_i\|_\infty, \|\beta_i\|_\infty, \, i = 1, 2\right\},$$

where

$$|f(t, \delta_1(t, u(t)), \delta_2(t, v(t)), u'(t), v'(t))| \le \psi_\rho(t), \text{ a.e. } t \in \mathbb{R}, \qquad (4.15)$$

$$|h(t, \delta_1(t, u(t)), \delta_2(t, v(t)), u'(t), v'(t))| \le \psi_\rho^*(t), \text{ a.e. } t \in \mathbb{R}. \qquad (4.16)$$

Define

$$M_i(s) := \max\left\{\sup_{t \in \mathbb{R}} |G_i(x, s)|, \sup_{t \in \mathbb{R}} \left|\frac{\partial G_i(t, s)}{\partial t}\right|, \, i = 1, 2\right\}, \qquad (4.17)$$

and remark that

$$\left|M_i(s)\right| \le \max\left\{\frac{1}{2}, \frac{1}{2k_i}, \, i = 1, 2\right\}, \, \forall s \in \mathbb{R}. \qquad (4.18)$$

For some

$$\rho_1 > \max\left\{\rho, \int_{-\infty}^{+\infty} M_1(s)\psi_\rho(s)ds, \int_{-\infty}^{+\infty} M_2(s)\psi_\rho^*(s)ds\right\}, \qquad (4.19)$$

let B be a bounded set of X, defined by

$$B := \{(u, v) \in X : \max\{\|u\|_\infty, \|u'\|_\infty, \|v\|_\infty, \|v'\|_\infty\} \le \rho_1\}. \qquad (4.20)$$

So,

$$\|T_1(u, v)(t)\|_\infty$$
$$= \sup_{t \in \mathbb{R}}\left(\left|\int_{-\infty}^{+\infty} G_1(t, s)f(s, \delta_1(s, u(s)), \delta_2(s, v(s)), u'(s), v'(s))ds\right|\right)$$
$$\le \int_{-\infty}^{+\infty} M_1(s)\psi_\rho(s)ds < +\infty,$$

and

$$\|(T_1(u, v))'(t)\|_\infty$$
$$= \sup_{t \in \mathbb{R}}\left(\left|\int_{-\infty}^{+\infty} \frac{\partial G_1(t, s)}{\partial t} f(s, \delta_1(s, u(s)), \delta_2(s, v(s)), u'(s), v'(s))ds\right|\right)$$
$$\le \int_{-\infty}^{+\infty} M_1(s)\psi_\rho(s)ds < +\infty.$$

So, T_1B is uniformly bounded on X_1. By similar arguments, we conclude that T_2 is also uniformly bounded on X_1. Therefore TB is uniformly bounded on X.

(iii) *TB is equicontinuous on X.*

Let $t_1, t_2 \in \mathbb{R}$ and suppose, without loss of generality, that $t_1 \leq t_2$. By the continuity of $G_1(t, s)$

$$\lim_{t_1 \to t_2} |T_1(u, v)(t_1) - T_1(u, v)(t_2)|$$

$$\leq \int_{-\infty}^{+\infty} \lim_{t_1 \to t_2} |G_1(t_1, s) - G_1(t_2, s)|$$

$$\times |f(s, \delta_1(s, u(s)), \delta_2(s, v(s)), u'(s), v'(s))| \, ds = 0.$$

As $\frac{\partial G_1(t,s)}{\partial t}$ is bounded on $\mathbb{R}^2$, therefore

$$\lim_{t_1 \to t_2} |(T_1(u, v))'(t_1) - (T_1(u, v))'(t_2)|$$

$$\leq \lim_{t_1 \to t_2} \int_{-\infty}^{+\infty} \left| \frac{\partial G_1(t_1, s)}{\partial t} - \frac{\partial G_1(t_2, s)}{\partial t} \right|$$

$$\times |f(s, \delta_1(s, u(s)), \delta_2(s, v(s)), u'(s), v'(s))| \, ds$$

$$= \lim_{t_1 \to t_2} \int_{-\infty}^{t_1} \left| \frac{\partial G_1(t_1, s)}{\partial t} - \frac{\partial G_1(t_2, s)}{\partial t} \right|$$

$$\times |f(s, \delta_1(s, u(s)), \delta_2(s, v(s)), u'(s), v'(s))| \, ds$$

$$+ \lim_{t_1 \to t_2} \int_{t_1}^{t_2} \left| \frac{\partial G_1(t_1, s)}{\partial t} - \frac{\partial G_1(t_2, s)}{\partial t} \right|$$

$$\times |f(s, \delta_1(s, u(s)), \delta_2(s, v(s)), u'(s), v'(s))| \, ds$$

$$+ \lim_{t_1 \to t_2} \int_{t_2}^{+\infty} \left| \frac{\partial G_1(t_1, s)}{\partial t} - \frac{\partial G_1(t_2, s)}{\partial t} \right|$$

$$\times |f(s, \delta_1(s, u(s)), \delta_2(s, v(s)), u'(s), v'(s))| \, ds.$$

By the continuity of $\frac{\partial G_1(t_1,s)}{\partial t}$, for $s \in \,]-\infty, t_1[$ and the continuity of $\frac{\partial G_1(t_2,s)}{\partial t}$ on $s \in \,]t_2, +\infty[$, then the first and the third integrals tend to 0. As the second one we have, by (4.15) and (4.18)

$$\lim_{t_1 \to t_2} \int_{t_1}^{t_2} \left| \frac{\partial G_1(t_1, s)}{\partial t} - \frac{\partial G_1(t_2, s)}{\partial t} \right|$$

$$\times |f(s, \delta_1(s, u(s)), \delta_2(s, v(s)), u'(s), v'(s))| \, ds = 0.$$

Therefore, T_1B is equicontinuous on X_1. Analogously, it can be proved that T_2B is equicontinuous on X_1 and so, TB is equicontinuous on X.

(iv) TD *is equiconvergent at* $t = \pm\infty$.

For the operator T_1, we have, by (i) and (4.15)

$$\left| T_1\left(u,v\right)(t) - \lim_{t\to\pm\infty} T_1\left(u,v\right)(t) \right|$$

$$\leq \int_{-\infty}^{+\infty} |G_1(t,s)|\,|f(s,\delta_1(s,u(s)),\delta_2(s,v(s)),u'(s),v'(s))|\,ds$$

$$\leq \int_{-\infty}^{+\infty} |G_1(t,s)|\,|\psi_\rho(s)|\,ds \to 0,$$

as $t \to \pm\infty$ and, by (4.18),

$$\left| (T_1\left(u,v\right))'(t) - \lim_{t\to\pm\infty} (T_1\left(u,v\right))'(t) \right|$$

$$\leq \int_{-\infty}^{+\infty} \left| \frac{\partial G_1(t,s)}{\partial t} \right| |f(s,\delta_1(s,u(s)),\delta_2(s,v(s)),u'(s),v'(s))|\,ds$$

$$\leq \int_{-\infty}^{+\infty} \left| \frac{\partial G_1(t,s)}{\partial t} \right| |\psi_\rho(s)|\,ds \to 0,$$

as $t \to \pm\infty$.

Therefore, T_1D is equiconvergent at $\pm\infty$ and, following a similar technique, we can prove that T_2D is equiconvergent at $\pm\infty$, too. So, TD is equiconvergent at $\pm\infty$.

(v) *For $D \subset X$ a nonempty, closed, bounded and convex set, we have* $T(D) \subset D$.

Let $\rho_2 > \rho_1$, with ρ_1 given by (4.19), and consider

$$D := \{(u,v) \in X : \|(u,v)\|_X \leq \rho_2\}.$$

Applying the same method as in Step **(ii)**, we have $\|T_1(u,v)\|_X \leq \rho_2$ and $\|T_2(u,v)\|_X \leq \rho_2$ and therefore $\|T(u,v)\|_X \leq \rho_2$.

By Theorem 3.1, TD is relatively compact, therefore, by Theorem 5.2, T has at least one fixed point $(u,v) \in X$, which is a solution of problem (4.12), (4.2).

(vi) *This solution of (4.12), (4.2), $(u,v) \in X$, is a solution of the initial problem (4.1), (4.2) if*

$$\alpha_1(t) \leq u(t) \leq \beta_1(t), \quad \alpha_2(t) \leq v(t) \leq \beta_2(t), \quad \forall t \in \mathbb{R}.$$

Let $(u(t), v(t))$ be a solution of problem (4.12), (4.2) and suppose, by contradiction, that there is $t \in \mathbb{R}$, such that $\alpha_1(t) > u(t)$. Define

$$\inf_{t \in \mathbb{R}} (u(t) - \alpha_1(t)) := u(t_0) - \alpha_1(t_0) < 0.$$

As, by (4.2) and Definition 4.1,

$$u(\pm\infty) - \alpha_1(\pm\infty) = -\alpha_1(\pm\infty) \geq 0,$$

therefore, $t_0 \neq \pm\infty$.

Assume that $t_0 \in \mathbb{R}$, such that

$$\min_{t \in \mathbb{R}} (u(t) - \alpha_1(t)) := u(t_0) - \alpha_1(t_0) < 0.$$

So, there exists an interval, $[t_1, t_2]$ such that $t_0 \in [t_1, t_2]$, and

$$u(t) - \alpha_1(t) < 0, \ u''(t) - \alpha_1''(t) \leq 0, \text{ a.e. } t \in [t_1, t_2],$$
$$u'(t) - \alpha_1'(t) \leq 0, \ t \in [t_1, t_0] \text{ and } u'(t) - \alpha_1'(t) \geq 0, \ t \in [t_0, t_2].$$

If $f(t, x, y, z, w)$ is nonincreasing on z, for $t \in [t_0, t_2]$, we have the following contradiction:

$$0 \leq \int_{t_0}^{t} \Big(u''(s) - \alpha_1''(s) \Big) ds$$
$$= \int_{t_0}^{t} \Big[f(s, \delta_1(s, u(s)), \delta_2(s, v(s)), u'(s), v'(s)) + k_1^2 u(s) - \alpha_1''(s) \Big] ds$$
$$\leq \int_{t_0}^{t} \Big[f(s, \alpha_1(s), \alpha_2(s), \alpha_1'(s), v'(s)) + k_1^2 u(s) - \alpha_1''(s) \Big] ds$$
$$\leq k_1^2 \int_{t_0}^{t} \Big[u(s) - \alpha_1(s) \Big] ds < 0.$$

If $f(t, x, y, z, w)$ is nondecreasing on z, we consider $t \in [t_1, t_0]$, and, from the previous arguments, we obtain a similar contradiction. Therefore, $\alpha_1(t) \leq u(t), \forall t \in \mathbb{R}$.

Applying the method above for function $h(t, x, y, z, w)$, it may be shown that $\alpha_2(t) \leq v(t) \leq \beta_2(t), \forall t \in \mathbb{R}$. $\square$

In Theorem 4.1, the guarantee that the homoclinic solution is non trivial is given by the assumptions

$$f(t, 0, 0, 0, 0) \neq 0, h(t, 0, 0, 0, 0) \neq 0, \text{ for some } t \in \mathbb{R}.$$

However, these conditions can be dropped, as it is shown in the following theorem, which proof is analogous to Theorem 4.1:

Theorem 4.2. *Let $f, h : \mathbb{R}^5 \to \mathbb{R}$ be L^1-Carathéodory functions verifying the assumptions of Theorem 4.1. Assume that there are (α_1, α_2), $(\beta_1, \beta_2) \in X$ lower and upper solutions of problem (4.1)-(4.2), respectively, such that*

$$0 < \alpha_i(t) \leq \beta_i(t), \text{ for } i = 1, 2 \text{ for some } t \in \mathbb{R},$$

or

$$\alpha_i(t) \leq \beta_i(t) < 0, \text{ for } i = 1, 2 \text{ for some } t \in \mathbb{R}.$$

If $f(t, x, y, z, w)$ is nonincreasing on y and monotone (nonincreasing or nondecreasing) on z, for fixed $(t, x, w) \in \mathbb{R}^3$, and $h(t, x, y, z, w)$ is nonincreasing in x and monotone (nonincreasing or nondecreasing) on w, for fixed $(t, y, z) \in \mathbb{R}^3$, then the problem (4.1)-(4.2) has a homoclinic solution $(u, v) \in X$ such that

$$\alpha_1(t) \leq u(t) \leq \beta_1(t), \quad \alpha_2(t) \leq v(t) \leq \beta_2(t), \ \forall t \in \mathbb{R}.$$

4.3 Application to a coupled nonlinear of real Schrödinger system type

In this section we consider an application to a family of a coupled stationary nonlinear Schrödinger system (NLS).

The Schrödinger equations are physically linked, with, for example, waves and solitons, [123, 146], nonlinear optics and photons, [149, 218].

In [40], the authors prove the existence of two different kinds of homoclinic solutions to the origin, describing solitary waves of physical relevance, using a system of two coupled nonlinear Schrödinger equations with inhomogeneous parameters, including a linear coupling.

In [62] (see also [48, 49]), the author considers the system of two coupled Schrödinger equations, modelling spatial solitons in crystals, or modeling pulse propagation in fibers, [218],

$$\begin{cases} i\dfrac{\partial u}{\partial t} + r\dfrac{\partial^2 u}{\partial x^2} - u + uv = 0 \\ i\dfrac{\partial v}{\partial t} + s\dfrac{\partial^2 v}{\partial x^2} - \alpha v + \dfrac{1}{2}u^2 = 0, \end{cases} \tag{4.21}$$

where:

- $u(t, x)$ and $v(t, x)$ represent the first and second harmonics of the amplitude envelope of an optical pulse, respectively;

- t and x are the time and space, respectively;
- α is the wave-vector mismatch between the two harmonics;
- $r, s = \pm 1$ and their signs are determined by the signs of the dispersions/diffractions (temporal/spatial cases, respectively).

In this application, we consider a real stationary family of system (4.21) (see [49]), with $r = s = \alpha = 1$ and $u(t)$, $v(t)$ are considered real,

$$
\begin{cases}
u''(t) - u(t) = -\dfrac{1}{9(t^2 + 1)} |u(t) - 1| \, (v(t) + 1) \\
v''(t) - v(t) = -\dfrac{1}{9(2t^2 + 1)} \, (u(t) + 1) \, |v(t) - 1|,
\end{cases}
\tag{4.22}
$$

together with the boundary conditions (4.2).

Notice that, the system (4.22), (4.2) is a particular case of problem (4.1), (4.2) with $k_1 = k_2 = 1$, and

$$
f(t, x, y, z, w) = -\frac{1}{9(t^2 + 1)} |x - 1| \, (y + 1),
$$

$$
h(t, x, y, z, w) = -\frac{1}{9(2t^2 + 1)} \, (x + 1) \, |y - 1|.
$$

Moreover, f and h verify the monotone assumptions of Theorem 4.1, and for $|x|, |y| < \rho$, both are L^1-Carathéodory functions, with

$$
\phi_\rho(t) = \frac{1}{9(t^2 + 1)} \, (\rho + 1)^2, \quad \varphi_\rho(t) = \frac{1}{9(2t^2 + 1)} \, (\rho + 1)^2.
$$

The functions $\alpha_i, \beta_i : \mathbb{R} \to \mathbb{R}$, $i = 1, 2$, given by $\alpha_i(t) = -1$ and $\beta_i(t) = 1$ are, respectively, lower and upper solutions of problem (4.22), (4.2), satisfying (4.10), once that

$$
\begin{aligned}
\alpha_1''(t) - \alpha_1(t) &= 1 > f(t, -1, -1, 0, w) = 0, \\
\alpha_2''(t) - \alpha_2(t) &= 1 > h(t, -1, -1, z, 0) = 0, \\
\beta_1''(t) - \beta_1(t) &= -1 < f(t, 1, 1, 0, w) = 0, \\
\beta_2''(t) - \beta_2(t) &= -1 < h(t, 1, 1, z, 0) = 0.
\end{aligned}
$$

As assumptions of Theorem 4.1 are verified for $\rho \in [1, \ 3.4]$, then the problem (4.22), (4.2) has a homoclinic solution $(u, v) \in X$ such that, for $i = 1, 2$,

$$
-1 \leq u(t) \leq 1 \ \text{and} \ -1 \leq v(t) \leq 1, \forall t \in \mathbb{R}.
$$

Remark that this solution is not the trivial one, as the null function is not a solution of (4.22).

Chapter 5

Heteroclinic solutions with phi-Laplacians

In this work, we consider the second-order coupled system on the real line

$$\begin{cases} \big(a(t)\phi\big(u'(t)\big)\big)' = f(t, u(t), v(t), u'(t), v'(t)), \\[2ex] \big(b(t)\psi\big(v'(t)\big)\big)' = h(t, u(t), v(t), u'(t), v'(t)), \ t \in \mathbb{R}, \end{cases} \tag{5.1}$$

with ϕ and ψ increasing homeomorphisms verifying some adequate relations on their inverses, $a, b : \mathbb{R} \to (0, +\infty[$ are continuous functions, $f, h : \mathbb{R}^5 \to \mathbb{R}$ are L^1-Carathéodory functions, together with asymptotic conditions

$$u(-\infty) = A, \ u'(+\infty) = 0, \ v(-\infty) = B, \ v'(+\infty) = 0 \tag{5.2}$$

for $A, B \in \mathbb{R}$.

Heteroclinic trajectories play an important role in the geometrical analysis of dynamical systems, connecting unstable and stable equilibria having two or more equilibrium points, [133]. In fact, the homoclinic or heteroclinic orbits are a kind of spiral structures, which are general phenomena in nature, [276]. Graphical illustrations and a very complete explanation on homoclinic and heteroclinic bifurcations can be seen in [122]. A planar homoclinic theorem and heteroclinic orbits, to analyze fluid models, is studied in [44]. Applications of dynamic systems techniques to the problem of heteroclinic connections and resonance transitions, are treated in [151], on planar circular domains. To prove the existence of heteroclinic solutions, for a class of non-autonomous second-order equations, see [15, 80, 184]. Topological, variational and minimization methods to find heteroclinic connections can be found in [270].

On heteroclinic coupled systems, among many published works, we highlight some of them:

In [8], Aguiar *et al.*, consider the dynamics of small networks of coupled cells, with one of the points, analyzed as invariant subsets, can support robust heteroclinic attractors;

In [23], Ashwin and Karabacak study coupled phase oscillators and discuss heteroclinic cycles and networks between partially synchronized states and in [143], they analyze coupled phase oscillators, highlighting a dynamic mechanism, nothing more than a heteroclinic network;

In [24], the authors investigate such heteroclinic network between partially synchronized states, where the phases cluster into three groups;

Moreover, in [91], the authors present some applications, results, methods and problems that have been recently reported and, in addition, they suggested some possible research directions, and some problems for further studies on homoclinics and heteroclinics.

Cabada and Cid, in [52], study the following boundary value problem on the real line

$$\begin{cases} \left(\phi\big(u'(t)\big)\right)' = f(t, u(t), u'(t)), \text{ on } \mathbb{R}, \\ u(-\infty) = -1, \quad u(+\infty) = 1, \end{cases}$$

with a singular ϕ-Laplacian operator where f is a continuous function that satisfies suitable symmetric conditions.

In [57], Calamai discusses the solvability of the following strongly nonlinear problem:

$$\begin{cases} \left(a(x(t))\phi\big(x'(t)\big)\right)' = f(t, x(t), x'(t)), \quad t \in \mathbb{R}, \\ x(-\infty) = \alpha, \quad x(+\infty) = \beta, \end{cases}$$

where $\alpha < \beta$, $\phi : (-r, r) \to \mathbb{R}$ is a general increasing homeomorphism with bounded domain (singular ϕ-Laplacian), a is a positive, continuous function and f is a Carathéodory nonlinear function.

Recently, in [140], Kajiwara proved the existence of a heteroclinic solution of the FitzHugh-Nagumo type reaction-diffusion system, under certain conditions on the heterogeneity.

Motivated by these works and applying the techniques suggested in [175, 193, 196, 198], we apply the fixed point theory, to obtain sufficient conditions for the existence of heteroclinic solutions of the coupled system (5.1), (5.2), assuming some adequate conditions on ϕ^{-1}, ψ^{-1}.

We emphasize that it is the first time where heteroclinic solutions for second-order coupled differential systems are considered for systems with full nonlinearities depending on both unknown functions and their first derivatives. An example illustrates the potentialities of our main result, and an application to coupled nonlinear systems of two degrees of freedom (2-DOF), shows the applicability of the main theorem.

5.1 Notations and preliminary results

Consider the following spaces

$$X := \left\{ x \in C^1(\mathbb{R}) : \lim_{|t| \to \infty} x^{(i)}(t) \in \mathbb{R}, \, i = 0, 1 \right\},$$

equipped with the norm

$$\|x\|_X = \max \left\{ \|x\|_\infty, \|x'\|_\infty \right\},$$

where

$$\|x\|_\infty := \sup_{t \in \mathbb{R}} |x(t)|,$$

and $X^2 := X \times X$ with

$$\|(u, v)\|_{X^2} = \max \left\{ \|u\|_X, \|v\|_X \right\}.$$

It can be proved that $(X, \|\cdot\|_X)$ and $(X^2, \|\cdot\|_{X^2})$ are Banach spaces.

Remark 5.1. If $w \in X$ then $w'(\pm\infty) = 0$.

By solution of problem (5.1), (5.2) we mean a pair $(u, v) \in X^2$ such that

$$a(t)\phi(u'(t)) \in W^{1,1}(\mathbb{R}) \text{ and } b(t)\psi(v'(t)) \in W^{1,1}(\mathbb{R}),$$

verifying (5.1), (5.2).

For the reader's convenience we consider the definition of L^1-Carathéodory functions:

Definition 5.1. A function $g : \mathbb{R}^5 \to \mathbb{R}$ is L^1-Carathéodory if

(i) for each $(x, y, z, w) \in \mathbb{R}^4$, $t \mapsto g(t, x, y, z, w)$ is measurable on $\mathbb{R}$;
(ii) for a.e. $t \in \mathbb{R}$, $(x, y, z, w) \mapsto g(t, x, y, z, w)$ is continuous on $\mathbb{R}^4$;
(iii) for each $\rho > 0$, there exists a positive function $\vartheta_\rho \in L^1(\mathbb{R})$ such that, whenever $x, y, z, w \in [-\rho, \rho]$, then

$$|g(t, x, y, z, w)| \leq \vartheta_\rho(t), \text{ a.e. } t \in \mathbb{R}. \tag{5.3}$$

Along this chapter we assume that

(H1) $\phi, \psi : \mathbb{R} \longrightarrow \mathbb{R}$ are increasing homeomorphisms such that
    a) $\phi(\mathbb{R}) = \mathbb{R}$, $\phi(0) = 0$, $\psi(\mathbb{R}) = \mathbb{R}$, $\psi(0) = 0$;
    b) $|\phi^{-1}(x)| \leq \phi^{-1}(|x|)$, $|\psi^{-1}(x)| \leq \psi^{-1}(|x|)$.

(H2) $a, b : \mathbb{R} \to (0, +\infty[$ are positive continuous functions such that

$$\lim_{t \to \pm\infty} \frac{1}{a(t)} \in \mathbb{R} \text{ and } \lim_{t \to \pm\infty} \frac{1}{b(t)} \in \mathbb{R}.$$

A convenient criterion for the compacity of the operators is given by next theorem:

Theorem 5.1 ([196], Theorem 2.3). *A set $M \subset X$ is relatively compact if the following conditions hold:*

(i) *both $\{t \to x(t) : x \in M\}$ and $\{t \to x'(t) : x \in M\}$ are uniformly bounded;*

(ii) *both $\{t \to x(t) : x \in M\}$ and $\{t \to x'(t) : x \in M\}$ are equicontinuous on any compact interval of $\mathbb{R}$;*

(iii) *both $\{t \to x(t) : x \in M\}$ and $\{t \to x'(t) : x \in M\}$ are equiconvergent at $\pm\infty$, that is, for any given $\epsilon > 0$, there exists $t_\epsilon > 0$ such that*

$$|f(t) - f(\pm\infty)| < \epsilon, \ |f'(t) - f'(\pm\infty)| < \epsilon, \forall |t| > t_\epsilon, f \in M.$$

The existence tool will be given by Schauder's fixed point theorem:

Theorem 5.2 ([269]). *Let Y be a nonempty, closed, bounded and convex subset of a Banach space X, and suppose that $P : Y \to Y$ is a compact operator. Then P has at least one fixed point in Y.*

5.2   Existence of heteroclinics

In this section we prove the existence for a pair of heteroclinic solutions to the coupled system (5.1), (5.2), for some constants $A, B \in \mathbb{R}$.

Theorem 5.3. *Let ϕ, $\psi : \mathbb{R} \to \mathbb{R}$ be increasing homeomorphisms and $a, b : \mathbb{R} \to (0, +\infty[$ continuous functions satisfying (H1) and (H2). Assume that $f, h : \mathbb{R}^5 \to \mathbb{R}$ are L^1-Carathéodory functions and there is $R > 0$ and $\vartheta_R, \theta_R \in L^1(\mathbb{R})$ such that*

$$\int_{-\infty}^{+\infty} \phi^{-1} \left(\frac{\int_{-\infty}^{+\infty} \vartheta_R(r)dr}{a(s)} \right) ds < +\infty, \tag{5.4}$$

$$\int_{-\infty}^{+\infty} \psi^{-1} \left(\frac{\int_{-\infty}^{+\infty} \theta_R(r)dr}{b(s)} \right) ds < +\infty, \tag{5.5}$$

with

$$\sup_{t \in \mathbb{R}} \phi^{-1} \left(\frac{\int_{-\infty}^{+\infty} \vartheta_R(r)dr}{a(t)} \right) ds < +\infty, \quad \sup_{t \in \mathbb{R}} \psi^{-1} \left(\frac{\int_{-\infty}^{+\infty} \theta_R(r)dr}{b(t)} \right) ds < +\infty,$$

$$|f(t, x, y, z, w)| \leq \vartheta_R(t), \tag{5.6}$$

$$|h(t, x, y, z, w)| \leq \theta_R(t), \tag{5.7}$$

whenever $x, y, z, w \in [-R, R]$.
Then for given $A, B \in \mathbb{R}$ *the problem (5.1), (5.2) has, at least, a pair of heteroclinic solutions* $(u, v) \in X^2$.

Proof. Define the operators $T_1 : X^2 \to X$, $T_2 : X^2 \to X$ and $T : X^2 \to X^2$ by

$$T(u, v) = (T_1(u, v), T_2(u, v)), \tag{5.8}$$

with

$$(T_1(u, v))(t) = \int_{-\infty}^{t} \phi^{-1} \left(\frac{\int_{-\infty}^{s} f(r, u(r), v(r), u'(r), v'(r))dr}{a(s)} \right) ds + A,$$

$$(T_2(u, v))(t) = \int_{-\infty}^{t} \psi^{-1} \left(\frac{\int_{-\infty}^{s} h(r, u(r), v(r), u'(r), v'(r))dr}{b(s)} \right) ds + B,$$

with A and B given by (5.2).

In order to apply Theorem 5.2, we shall prove that T is compact and has a fixed point.

To simplify the proof, we detail the arguments only for $T_1(u, v)$, as for the operator $T_2(u, v)$ the technique is similar.

To be clear, we divide the proof into claims **(i)-(v)**.

(i) T *is well defined and continuous in* X^2.

Let $(u, v) \in X^2$ and take $\rho > 0$ such that $\|(u, v)\|_{X^2} < \rho$. As f is a L^1-Carathéodory function, there exists a positive function $\vartheta_\rho \in L^1(\mathbb{R})$ verifying (5.6). So,

$$\int_{-\infty}^{t} |f(r, u(r), v(r), u'(r), v'(r))|dr$$

$$\leq \int_{-\infty}^{+\infty} |f(r, u(r), v(r), u'(r), v'(r))|dr \leq \int_{-\infty}^{+\infty} \vartheta_\rho(t)dt < +\infty.$$

So, T_1 is continuous on X. Furthermore,

$$(T_1(u, v))'(t) = \phi^{-1} \left(\frac{\int_{-\infty}^{t} f(r, u(r), v(r), u'(r), v'(r))dr}{a(t)} \right)$$

is also continuous on X and, therefore, $T_1(u, v) \in C^1(\mathbb{R})$.

By (5.2), (5.4), (5.6) and (H2),

$$\lim_{t \to -\infty} T_1(u,v)(t)$$

$$= \lim_{t \to -\infty} \int_{-\infty}^{t} \phi^{-1} \left(\frac{\int_{-\infty}^{s} f(r, u(r), v(r), u'(r), v'(r)) dr}{a(s)} \right) ds + A = A,$$

$$\lim_{t \to +\infty} T_1(u,v)(t)$$

$$= \int_{-\infty}^{+\infty} \phi^{-1} \left(\frac{\int_{-\infty}^{s} f(r, u(r), v(r), u'(r), v'(r)) dr}{a(s)} \right) ds + A < +\infty,$$

and

$$\lim_{t \to \pm\infty} (T_1(u,v)(t))' = \lim_{t \to \pm\infty} \phi^{-1} \left(\frac{\int_{-\infty}^{t} f(r, u(r), v(r), u'(r), v'(r)) dr}{a(t)} \right)$$

$$\leq \lim_{t \to \pm\infty} \phi^{-1} \left(\frac{\int_{-\infty}^{+\infty} \vartheta_\rho(r) dr}{a(t)} \right) < +\infty.$$

Therefore, $T_1(u,v) \in X$, and, by the same arguments, $T_2(u,v) \in X$. So, $T(u,v) \in X^2$.

(ii) TM *is uniformly bounded on* $M \subseteq X^2$, *for some bounded* M.

Let M be a bounded set of X^2, defined by

$$M := \{(u,v) \in X^2 : \max\{\|u\|_\infty, \|u'\|_\infty, \|v\|_\infty, \|v'\|_\infty\} \leq \rho_1\}, \qquad (5.9)$$

for some $\rho_1 > 0$.
By (5.4), (5.6), (H1) and (H2), we have

$$\|T_1(u,v)(t)\|_\infty$$

$$= \sup_{t \in \mathbb{R}} \left(\left| \int_{-\infty}^{t} \phi^{-1} \left(\frac{\int_{-\infty}^{s} f(r, u(r), v(r), u'(r), v'(r)) dr}{a(s)} \right) ds + A \right| \right.$$

$$\leq \sup_{t \in \mathbb{R}} \int_{-\infty}^{t} \left| \phi^{-1} \left(\frac{\int_{-\infty}^{s} f(r, u(r), v(r), u'(r), v'(r)) dr}{a(s)} \right) \right| ds + |A|$$

$$\leq \sup_{t \in \mathbb{R}} \int_{-\infty}^{t} \phi^{-1} \left(\frac{\int_{-\infty}^{s} |f(r, u(r), v(r), u'(r), v'(r))| dr}{a(s)} \right) ds + |A|$$

$$\leq \int_{-\infty}^{+\infty} \phi^{-1} \left(\frac{\int_{-\infty}^{+\infty} \vartheta_{\rho_1}(r) dr}{a(s)} \right) ds + |A| < +\infty,$$

and

$$\| (T_1 (u, v))' (t)\|_\infty$$

$$= \sup_{t \in \mathbb{R}} \left| \phi^{-1} \left(\frac{\int_{-\infty}^{t} f(r, u(r), v(r), u'(r), v'(r)) dr}{a(t)} \right) \right|$$

$$\leq \sup_{t \in \mathbb{R}} \phi^{-1} \left(\frac{\int_{-\infty}^{t} |f(r, u(r), v(r), u'(r), v'(r))| \, dr}{a(t)} \right)$$

$$\leq \sup_{t \in \mathbb{R}} \phi^{-1} \left(\frac{\int_{-\infty}^{+\infty} \vartheta_{\rho_1}(r) dr}{a(t)} \right) < +\infty.$$

So, $\|T_1 (u, v) (t)\|_X < +\infty$, that is, $T_1 M$ is uniformly bounded on X.

By similar arguments, T_2 is uniformly bounded on X. Therefore TM is uniformly bounded on X^2.

(iii) TM *is equicontinuous on* X^2.

Let $t_1, t_2 \in [-K, K] \subseteq \mathbb{R}$ for some $K > 0$, and suppose, without loss of generality, that $t_1 \leq t_2$. Thus, by (5.4), (5.6) and (H1),

$$|T_1 (u, v) (t_1) - T_1 (u, v) (t_2)|$$

$$= \left| \int_{-\infty}^{t_1} \phi^{-1} \left(\frac{\int_{-\infty}^{s} f(r, u(r), v(r), u'(r), v'(r)) dr}{a(s)} \right) ds \right.$$

$$\left. - \int_{-\infty}^{t_2} \phi^{-1} \left(\frac{\int_{-\infty}^{s} f(r, u(r), v(r), u'(r), v'(r)) dr}{a(s)} \right) ds \right|$$

$$\leq \int_{t_1}^{t_2} \phi^{-1} \left(\frac{\int_{-\infty}^{s} |f(r, u(r), v(r), u'(r), v'(r)) dr|}{a(s)} \right) ds$$

$$\leq \int_{t_1}^{t_2} \phi^{-1} \left(\frac{\int_{-\infty}^{+\infty} \vartheta_{\rho_1}(r) dr}{a(s)} \right) ds \to 0,$$

uniformly for $(u, v) \in M$, as $t_1 \to t_2$, and

$$\left| (T_1 (u, v))' (t_1) - (T_1 (u, v))' (t_2) \right|$$

$$= \left| \phi^{-1} \left(\frac{\int_{-\infty}^{t_1} f(r, u(r), v(r), u'(r), v'(r)) dr}{a(t_1)} \right) \right.$$

$$\left. - \phi^{-1} \left(\frac{\int_{-\infty}^{t_2} f(r, u(r), v(r), u'(r), v'(r)) dr}{a(t_2)} \right) \right| \to 0,$$

uniformly for $(u, v) \in M$, as $t_1 \to t_2$.

Therefore, $T_1 M$ is equicontinuous on X. Analogously, it can be proved that $T_2 M$ is equicontinuous on X. So, TM is equicontinuous on X^2.

(iv) TM *is equiconvergent at* $t = \pm\infty$.

Let $(u, v) \in M$. For the operator T_1, we have, by (5.4), (5.6) and (H1),

$$\left| T_1 (u, v) (t) - \lim_{t \to -\infty} T_1 (u, v) (t) \right|$$

$$= \left| \int_{-\infty}^{t} \phi^{-1} \left(\frac{\int_{-\infty}^{s} f(r, u(r), v(r), u'(r), v'(r)) dr}{a(s)} \right) ds \right|$$

$$\leq \int_{-\infty}^{t} \phi^{-1} \left(\frac{\int_{-\infty}^{+\infty} \vartheta_{\rho_1}(r) dr}{a(s)} \right) ds \to 0,$$

uniformly in $(u, v) \in M$, as $t \to -\infty$, and,

$$\left| T_1 (u, v) (t) - \lim_{t \to +\infty} T_1 (u, v) (t) \right|$$

$$= \left| \int_{-\infty}^{t} \phi^{-1} \left(\frac{\int_{-\infty}^{s} f(r, u(r), v(r), u'(r), v'(r)) dr}{a(s)} \right) ds \right.$$

$$\left. - \int_{-\infty}^{+\infty} \phi^{-1} \left(\frac{\int_{-\infty}^{s} f(r, u(r), v(r), u'(r), v'(r)) dr}{a(s)} \right) ds \right|$$

$$= \left| \int_{t}^{+\infty} \phi^{-1} \left(\frac{\int_{-\infty}^{s} f(r, u(r), v(r), u'(r), v'(r)) dr}{a(s)} \right) ds \right|$$

$$\leq \int_{t}^{+\infty} \phi^{-1} \left(\frac{\int_{-\infty}^{+\infty} \vartheta_{\rho_1}(r) dr}{a(s)} \right) ds \to 0,$$

uniformly in $(u, v) \in M$, as $t \to +\infty$.
For the derivative it follows that,

$$\left| (T_1 (u, v))' (t) - \lim_{t \to +\infty} (T_1 (u, v))' (t) \right|$$

$$= \left| \phi^{-1} \left(\frac{\int_{-\infty}^{t} f(r, u(r), v(r), u'(r), v'(r)) dr}{a(t)} \right) \right.$$

$$\left. - \phi^{-1} \left(\lim_{t \to +\infty} \frac{\int_{-\infty}^{+\infty} f(r, u(r), v(r), u'(r), v'(r)) dr}{a(t)} \right) \right| \to 0,$$

uniformly in $(u, v) \in M$, as $t \to +\infty$, and

$$\left| (T_1(u, v))'(t) - \lim_{t \to -\infty} (T_1(u, v))'(t) \right|$$

$$= \left| \phi^{-1} \left(\frac{\int_{-\infty}^{t} f(r, u(r), v(r), u'(r), v'(r)) dr}{a(t)} \right) \right|$$

$$\leq \phi^{-1} \left(\left| \frac{\int_{-\infty}^{t} f(r, u(r), v(r), u'(r), v'(r)) dr}{a(t)} \right| \right)$$

$$\leq \phi^{-1} \left(\frac{\int_{-\infty}^{t} \vartheta_{\rho_1}(r) dr}{|a(t)|} \right) \to 0$$

uniformly in $(u, v) \in M$, as $t \to -\infty$.

Therefore, $T_1 M$ is equiconvergent at $\pm\infty$ and, following a similar technique, we can prove that $T_2 M$ is equiconvergent at $\pm\infty$, too. So, TM is equiconvergent at $\pm\infty$.

By Theorem 5.1, TM is relatively compact.

(v) $T : X \to X$ *has a fixed point.*

In order to apply Schauder's fixed point theorem for operator $T(u, v)$, we need to prove that $TD \subset D$, for some closed, bounded and convex $D \subset X^2$.

Consider

$$D := \left\{ (u, v) \in X^2 : \| (u, v) \|_{X^2} \leq \rho_2 \right\},$$

with $\rho_2 > 0$ such that

$$\rho_2 := \max \left\{ \begin{array}{l} \rho_1, \ \int_{-\infty}^{+\infty} \phi^{-1} \left(\frac{\int_{-\infty}^{+\infty} \vartheta_{\rho_2}(r) dr}{a(s)} \right) ds + |A|, \\[3mm] \int_{-\infty}^{+\infty} \psi^{-1} \left(\frac{\int_{-\infty}^{+\infty} \theta_{\rho_2}(r) dr}{b(s)} \right) ds + |B|, \\[3mm] \sup_{t \in \mathbb{R}} \phi^{-1} \left(\frac{\int_{-\infty}^{+\infty} \vartheta_{\rho_2}(r) dr}{a(t)} \right), \ \sup_{t \in \mathbb{R}} \psi^{-1} \left(\frac{\int_{-\infty}^{+\infty} \theta_{\rho_2}(r) dr}{b(t)} \right) \end{array} \right\}$$

with ρ_1 given by (5.9).

Following similar arguments as in **(ii)**, we have, for $(u, v) \in D$,

$$\| T(u, v) \|_{X^2} = \| (T_1(u, v), T_2(u, v)) \|_{X^2}$$

$$= \max \left\{ \| T_1(u, v) \|_X, \| T_2(u, v) \|_X \right\}$$

$$= \max \left\{ \begin{array}{l} \| T_1(u, v) \|_\infty, \ \| (T_1(u, v))' \|_\infty, \\ \| T_2(u, v) \|_\infty, \ \| (T_2(u, v))' \|_\infty \end{array} \right\} \leq \rho_2,$$

and $TD \subset D$.

By Theorem 5.2, the operator $T(u, v) = (T_1(u, v), T_2(u, v))$ has a fixed point $(u, v) \in X^2$.

By standard arguments, it can be proved that this fixed point defines a pair of heteroclinic or homoclinic solutions of problem (5.1), (5.2).   □

Remark 5.2. If

$$\int_{-\infty}^{+\infty} \phi^{-1} \left(\frac{\int_{-\infty}^{s} f(r, u(r), v(r), u'(r), v'(r)) dr}{a(s)} \right) ds = 0$$

and

$$\int_{-\infty}^{+\infty} \psi^{-1} \left(\frac{\int_{-\infty}^{s} h(r, u(r), v(r), u'(r), v'(r)) dr}{b(s)} \right) ds = 0,$$

the solutions $(u, v) \in X^2$ of problem (5.1), (5.2), will be a pair of homoclinic solutions.

5.3 Application to coupled systems of nonlinear 2-DOF model

Generic nonlinear coupled systems of two degrees of freedom (2-DOF), are especially important in Physics and Mechanics. For example in [188], the authors use this type of system to investigate the transient in a system containing a linear oscillator, linearly coupled to an essentially nonlinear attachment with a comparatively small mass. The family of coupled nonlinear systems of 2-DOF is used to study the global bifurcations in the motion of an externally forced coupled nonlinear oscillatory system or for the nonlinear vibration absorber subjected to periodic excitation, see [183]. Moreover, in [20], the authors deal with the stochastic moment stability of such systems.

Motivated by these works, in this section we consider an application of system (5.1), (5.2), to a family of coupled nonlinear systems of 2-DOF model, given by the nonlinear coupled system (see [183])

$$\begin{cases} \left((1 + t^4)(q_1'(t))^3 \right)' = \frac{t^4}{(1+t^6)^2} \left[2\zeta\omega_0 (q_1'(t))^3 + \omega_0^2 q_1(t) + \gamma((q_1(t))^3 \right. \\ \left. \qquad -3d^2 q_1(t)q_2(t)) + \cos(t) \right], \\ \\ \tau^2 \left((1 + t^4)(q_2'(t))^3 \right)' = \frac{t^4}{(1+t^6)^2} \left[2\zeta\omega_0 (q_2'(t))^3 + \omega_0^2 q_2(t) + \gamma(d^2 (q_2(t))^3 \right. \\ \left. \qquad -3 (q_1(t))^2 q_2(t)) \right], \ t \in \mathbb{R}, \end{cases}$$

$$\tag{5.10}$$

where

- $q_1(t)$ and $q_2(t)$ represent the generalized coordinates;
- d, τ, γ are positive constant coefficients which depend on the characteristics of the physical or mechanical system under consideration;
- $\cos(t)$ is related to the type of excitation of the system under consideration;
- ζ, ω_0, are the damping coefficient and the frequency, respectively.

As the asymptotic conditions we consider

$$q_1(-\infty) = A, \quad q_1'(+\infty) = 0, \quad q_2(-\infty) = B, \quad q_2'(+\infty) = 0, \qquad (5.11)$$

with $A, B \in \mathbb{R}$, and, moreover, assume that the real coefficients ζ, ω_0, γ, d, r are such that the integrals

$$\int_{-\infty}^{+\infty} \left(\sqrt[3]{\frac{\int_{-\infty}^{s} \frac{r^4}{(1+r^6)^2} \left[\begin{array}{c} 2\zeta\omega_0 \left(q_1'(r)\right)^3 + \omega_0^2 q_1(r) + \gamma((q_1(r))^3 \\ -3d^2 q_1(r) q_2(r)) + \cos(r) \end{array} \right] dr}{1 + s^4}} \right) ds \tag{5.12}$$

and

$$\int_{-\infty}^{+\infty} \left(\sqrt[3]{\frac{\int_{-\infty}^{s} \frac{r^4}{\tau^2(1+r^6)^2} \left[\begin{array}{c} 2\zeta\omega_0 \left(q_2'(r)\right)^3 + \omega_0^2 q_2(r) + \gamma(d^2 \left(q_2(r)\right)^3 \\ -3 \left(q_1(r)\right)^2 q_2(r)) \end{array} \right] dr}{1 + s^4}} \right) ds \tag{5.13}$$

are finite.

It is clear that (5.10) is a particular case of (5.1) with:

$$\phi(z) = \psi(z) = z^3, \quad a(t) = b(t) = 1 + t^4,$$

$f, h : \mathbb{R}^5 \to \mathbb{R}$ are L^1-Carathéodory functions where

$$
\begin{aligned}
f(t, x, y, z, w) &= \frac{t^4}{(1 + t^6)^2} \left(2\zeta\omega_0 z^3 + \omega_0^2 x + \gamma x^3 - 3d^2 xy + \cos(t) \right) \\
&\leq \frac{t^4}{(t^6 + 1)^2} \left(2 |\zeta\omega_0| \rho^3 + \omega_0^2 \rho + \gamma\rho^3 + 3d^2 \rho^2 + 1 \right) \\
&:= \delta_\rho(t)
\end{aligned}
$$

$$h(t, x, y, z, w) = \frac{t^4}{\tau^2(1+t^6)^2} \left(2\zeta\omega_0 w^3 + \omega_0^2 y + \gamma d^2 y^3 - 3x^2 y\right)$$

$$\leq \frac{t^4}{\tau^2(t^6+1)^2} \left(2|\zeta\omega_0|\rho^3 + \omega_0^2\rho + \gamma d^2\rho^3 + 3\rho^3\right)$$

$$:= \varepsilon_\rho(t)$$

where $\delta_\rho(t)$ and $\varepsilon_\rho(t)$ are functions in $L^1(\mathbb{R})$, for $\rho > 0$ such that

$$\rho := \max\{|x|, |y|, |z|, |w|\}. \tag{5.14}$$

Moreover, conditions (H1) and (H2) hold as,

- $\phi(\mathbb{R}) = \psi(w) = \mathbb{R}$ and $\phi(0) = \psi(0) = 0$;
- $|\phi^{-1}(z)| = |\sqrt[3]{z}| = \phi^{-1}(|z|) = \sqrt[3]{|z|}$ and $|\psi^{-1}(w)| = |\sqrt[3]{w}| = \psi^{-1}(|w|) = \sqrt[3]{|w|}$;
- $\lim_{t\to\pm\infty} \frac{1}{a(t)} = \lim_{t\to\pm\infty} \frac{1}{1+t^4} = \lim_{t\to\pm\infty} \frac{1}{b(t)} = 0$.

For $\rho > 0$ such that

$$\int_{-\infty}^{+\infty} \phi^{-1}\left(\frac{\int_{-\infty}^{+\infty} \delta_\rho(r)dr}{a(s)}\right) ds$$

$$= \int_{-\infty}^{+\infty} \left(\sqrt[3]{\frac{\int_{-\infty}^{+\infty} r^4 \frac{(2|\zeta\omega_0|\rho^3 + \omega_0^2\rho + \gamma\rho^3 + 3d^2\rho^2 + 1)}{(1+r^6)^2}dr}{1+s^4}}\right) ds < \rho \tag{5.15}$$

and

$$\int_{-\infty}^{+\infty} \psi^{-1}\left(\frac{\int_{-\infty}^{+\infty} \varepsilon_\rho(r)dr}{b(s)}\right) ds$$

$$= \int_{-\infty}^{+\infty} \left(\sqrt[3]{\frac{\int_{-\infty}^{+\infty} r^4 \frac{(2|\zeta\omega_0|\rho^3 + \omega_0^2\rho + \gamma d^2\rho^3 + 3\rho^3)}{\tau^2(1+r^6)^2}dr}{1+s^4}}\right) ds < \rho, \tag{5.16}$$

by Theorem 5.3, the system (5.10) together with the asymptotic conditions (5.11), has at least a pair $(q_1, q_2) \in X^2$ of heteroclinic solutions, since the integrals (5.12) and (5.13) are finite. As example, in particular, for

$$|\zeta| = \frac{1}{2\sqrt{1000}}, \quad |\omega_0| = \frac{1}{\sqrt{1000}}, \quad \gamma = \frac{1}{1000}, \quad d^2 = \frac{1}{3000}, \quad \tau = 23$$

the conditions (5.15) and (5.16) hold for $\rho > 6.3542$.

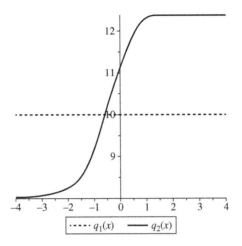

Fig. 5.1 $A = 10, B = 8$.

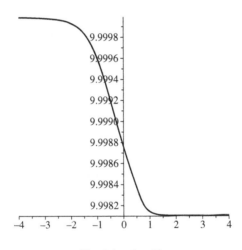

Fig. 5.2 $A = 10$.

For the values of the above parameters, $A = 10$ and $B = 8$, the hetero-clinic solutions q_1 and q_2 have the graphs given in Figure 5.1. In Figure 5.2 we present the real shape of the q_1 trajectory, which is not detailed in Figure 5.1 due to the scale range.

Remark that, if the integrals (5.12) and (5.13) are null, then the system (5.10) has a pair of homoclinic solutions $(q_1, q_2) \in X^2$.

Conclusions and open issues

The main goal of Part II is to provide some techniques to overcome the lack of compactness of the operators when the domain is unbounded.

In short, in Chapter 3 we combine an adequate Nagumo-type condition, to control the growth of the first derivatives, with the equiconvergence or the stability of the operator at $+\infty$, to apply the fixed point theory.

In Chapters 4 and 5, the problems are considered in all the real line, and the boundary restrictions are given as asymptotic conditions. In the first case, the existence and the localization of the homoclinic solutions are obtained for L^1-Carathéodory nonlinearities in the presence of well ordered lower and upper solutions. The last chapter deals with strongly nonlinear fully differential equations where the usual growth and/or assumptions on the nonlinearities are replaced by conditions on the homeomorphism to guarantee the existence of heteroclinic solutions for the coupled system.

Besides some of the open questions in the Conclusion of Part I remain applicable to coupled systems on unbounded intervals, there are new issues to be investigated forward:

- Can the problem be considered in a different functional framework, without the weighted norms, requiring less regularity to the nonlinearities?
- Is it possible to remove condition (3.11), or replaced it by a unilateral asymptotic condition near 0 or at ∞?
- Does a second-order coupled system with more general linear parts, defined on $\mathbb{R}$, have homoclinic solutions?

- What is the relation, if any, between the regularity of the kernel functions of the coupled system and the existence of homoclinic solutions?
- If, in Chapter 5, the assumption (H1) does not hold, what are the sufficient conditions for having heteroclinic solutions to the corresponding coupled system?

PART III
Coupled impulsive systems

Introduction

Impulsive differential equations describe processes in which a sudden change of state occurs at certain moments. Usually one of the characteristics of these processes is the existence of instantaneous disturbances or at a very short time in relation to the process itself.

These situations arise naturally, for example, in phenomena studied in physics, chemistry, population dynamics, biotechnology, economics, optimal control, medicine and others [29, 32, 59, 79, 124, 243, 263]. However, the study of systems involving two or more differential impulsive equations is scarce in both bounded and unbounded domains (see, [173]).

The first appearance of equations or systems involving impulses arises in the early 20th century and appears to be related to the Dirac delta distribution, [34, 65, 77].

To the best of our knowledge, the first paper to make reference on impulses is [190], about the stability of motion with the presence of impulses.

Since then, several different approaches and applications on systems, boundary value problems and numerical methods involving impulses were developed, [30, 31, 42, 75, 83, 85, 96, 189, 226, 231].

In this Part we study problems composed by second-order coupled systems with complete nonlinearities together with mixed boundary conditions at two points, and subject to generalized impulse conditions which allow jumps in the unknown functions and its derivative. Our goal, is to determine sufficient conditions on the nonlinearities to ensure the solvability of the problem.

To investigate this Part, we highlight the arguments used in [194, 195, 204], combined mainly with:

- *Carathéodory sequences* to control the behavior of the infinite moments of impulse;

- **Equiconvergence** at each impulsive moment and at $\pm\infty$ to recover the compacity on unbounded domains;
- Similar methods and techniques to those ones used in Parts I and II.

This third Part consists of three chapters which cover the existence and location of solutions for impulsive coupled systems on bounded and unbounded domains. More precisely:

- In *Chapter 6*, we consider impulsive coupled system with mixed boundary conditions, together with generalized impulsive conditions, with dependence on the first derivative on bounded intervals;
- In *Chapter 7*, we study the existence of solutions for impulsive coupled system with boundary conditions on unbounded intervals;
- Finally, in *Chapter 8*, we present some techniques to localize the solution, given in the previous chapter, applying the method of lower and upper solutions.

It is pointed out that, each chapter contains an application to real phenomena, to illustrate the applicability of our results in impulsive environments.

Chapter 6

Impulsive coupled systems with generalized jump conditions

In this chapter we consider the second-order impulsive coupled system with mixed boundary conditions

$$\begin{cases} u''(x) = f(x, u(x), u'(x), v(x), v'(x)) \\ v''(x) = h(x, u(x), u'(x), v(x), v'(x)) \\ \qquad u(a) = A_1, \ u'(b) = B_1, \\ \qquad v(a) = A_2, \ v'(b) = B_2, \end{cases} \tag{6.1}$$

with $f, h : [a, b] \times \mathbb{R}^4 \to \mathbb{R}$ L^1-Carathéodory functions, $A_1, A_2, B_1, B_2 \in \mathbb{R}$, with the generalized impulsive conditions

$$\begin{cases} \Delta u(x_k) = I_{0k}(x_k, u(x_k), u'(x_k)), \ k = 1, 2, \ldots, n, \\ \Delta u'(x_k) = I_{1k}(x_k, u(x_k), u'(x_k)), \\ \Delta v(\tau_j) = J_{0j}(\tau_j, v(\tau_j), v'(\tau_j)), \ j = 1, 2, \ldots, m, \\ \Delta v'(\tau_j) = J_{1j}(\tau_j, v(\tau_j), v'(\tau_j)), \end{cases} \tag{6.2}$$

where, for $i = 0, 1$, $\Delta u^{(i)}(x_k) = u^{(i)}(x_k^+) - u^{(i)}(x_k^-)$, $\Delta v^{(i)}(\tau_j) = v^{(i)}(\tau_j^+) - v^{(i)}(\tau_j^-)$, and being $I_{ik}, J_{ij} \in C([a, b] \times \mathbb{R}^2, \mathbb{R})$, with x_k, τ_j fixed points such that $a < x_1 < x_2 < \cdots < x_n < b$ and $a < \tau_1 < \tau_2 < \cdots < \tau_m < b$.

This chapter is based on [208].

The theory of impulsive differential equations describes processes in which a sudden change of state occurs at certain moments. Several authors (see for example, [2, 41, 92, 93, 104, 129, 130, 137, 165, 166, 177, 194, 203]) have dealt with impulsive differential equations, from different points of view and using many techniques.

There are many phenomena and applications related to impulsive differential systems, for example, we can find biological models, population dynamics, neural networks, models in economics, on time scales, on state-dependent delays, on delay-dependent impulsive control, on electrochemical communication between cells in the brain, (see for instance, $[42, 59, 89, 127, 167, 168, 181, 200, 202, 205, 213, 229, 243, 247]$), among others.

In [234], the author considers sufficient conditions for the existence and uniqueness of solutions to the following complex dynamical network in the form of a coupled system of $m + 2$ point boundary conditions for impulsive fractional differential equations

$$
\begin{cases}
{}^cD^\alpha u(t) = \phi(t, u(t), v(t)), \ \ t \in [0,1], \ \ t \neq t_j, \ j = 1, \ldots, m, \\[2mm]
{}^cD^\beta v(t) = \psi(t, u(t), v(t)), \ \ t \in [0,1], \ \ t \neq t_i, \ i = 1, \ldots, n, \\[2mm]
u(0) = h(u), \ \ u(1) = g(u) \ \text{and} \ v(0) = k(v), \ \ v(1) = f(v), \\[2mm]
\Delta u(t_j) = I_j(u(t_j)), \ \ \Delta u'(t_j) = \bar{I}_j(u(t_j)), \ \ j = 1, \ldots, m, \\[2mm]
\Delta v(t_i) = I_i(v(t_i)), \ \ \Delta v'(t_i) = \bar{I}_i(v(t_i)), \ \ i = 1, \ldots, n,
\end{cases}
$$

where $1 < \alpha, \beta \leq 2$, $\phi, \psi : [0,1] \times \mathbb{R}^2 \to \mathbb{R}$ are continuous functions and $g, h : X \to \mathbb{R}$, $f, k : Y \to \mathbb{R}$ are continuous functionals define by

$$
g(u) = \sum_{j=1}^{p} \lambda_j u(\xi_j), \ \ h(u) = \sum_{j=1}^{p} \lambda_j u(\eta_j),
$$

$$
f(v) = \sum_{i=1}^{q} \delta_i v(\xi_i), \ \ k(v) = \sum_{i=1}^{q} \delta_i v(\eta_i),
$$

$\xi_i, \eta_i, \xi_j, \eta_j \in (0,1)$ for $i = 1, ..., q$ and $j = 1, ..., q$.

In [173], the author studied the BVP composed by the second-order singular differential system on the whole line, with impulse effects, i.e., consisting of the differential system

$$
\begin{aligned}
[\phi_p(\rho(t)x'(t))]' &= f(t, x(t), y(t)), \ \ \text{a.e. } t \in \mathbb{R}, \\
[\phi_q(\varrho(t)y'(t))]' &= g(t, x(t), y(t)), \ \ \text{a.e. } t \in \mathbb{R}
\end{aligned}
$$

subjected to the asymptotic conditions

$$
\begin{aligned}
\lim_{t \to \pm\infty} x(s) &= 0, \\
\lim_{t \to \pm\infty} y(s) &= 0
\end{aligned}
$$

and the impulsive effects

$$\Delta x(t_k) = I_k(t_k, x(t_k), y(t_k)), \quad k \in \mathbb{Z}$$
$$\Delta y(t_k) = J_k(t_k, x(t_k), y(t_k)), \quad k \in \mathbb{Z},$$

where

(a) $\rho, \varrho \in C^0(\mathbb{R}, [0, \infty))$, $\rho(t), \varrho(t) > 0$ for all $t \in \mathbb{R}$ with $\int_{-\infty}^{+\infty} \frac{ds}{\rho(s)} < +\infty$
and $\int_{-\infty}^{+\infty} \frac{ds}{\varrho(s)} < +\infty$,

(b) $\phi_p(x) = x|x|^{p-2}$, $\phi_q(x) = x|x|^{q-2}$ with $p > 1$ and $q > 1$ are Laplace operators,

(c) f, g on $\mathbb{R}^3$ are Carathéodory functions,

(d) $\cdots < t_k < t_{k+1} < t_{k+2} < \cdots$ with $\lim_{k \to -\infty} t_k = -\infty$ and $\lim_{k \to +\infty} t_k = +\infty$, $\Delta x(t_k) = u(t_x^+) - x(t_k^-)$ and $\Delta y(t_k) = y(t_x^+) - y(t_k^-)$ ($k \in \mathbb{Z}$), $\mathbb{Z}$ is the set of all integers,

(c) $\{I_k\}, \{J_k\}$, with $I_k, J_k : \mathbb{R}^3 \to \mathbb{R}$ are Carathéodory sequences.

Motivated by these works, we follow arguments applied in [204], to study problem (6.1)-(6.2). We point out, that is the first time where second-order coupled systems include full nonlinearities. That is, they depend on the unknown functions and on their first derivatives, together with generalized impulsive conditions with dependence on the first derivative, too.

6.1 Definitions and auxiliary results

Define $u(x_k^\pm) := \lim_{x \to x_k^\pm} u(x)$ and consider the set

$$PC_1([a, b]) = \left\{ \begin{matrix} u : u \in C([a, b], \mathbb{R}) \text{ continuous for } x \neq x_k, u(x_k) = u(x_k^-), \\ u(x_k^+) \text{ exists for } k = 1, 2, \dots, n \end{matrix} \right\},$$

and the space $X_1 := PC_1^1([a, b]) = \{u : u'(t) \in PC_1([a, b])\}$ equipped with the norm $\|u\|_{X_1} = \max\{\|u\|, \|u'\|\}$, where

$$\|w\| := \sup_{x \in [a, b]} |w(x)|.$$

Analogously, define the set $X_2 := PC_2^1([a, b]) = \{v : v'(x) \in PC_2([a, b])\}$, with

$$PC_2([a, b]) = \left\{ \begin{matrix} v : v \in C([a, b], \mathbb{R}) \text{ continuous for } \tau \neq \tau_j, v(\tau_j) = v(\tau_j^-), \\ v(\tau_j^+) \text{ exists for } j = 1, 2, \dots, m \end{matrix} \right\},$$

equipped with the norm $\|v\|_{X_2} = \max\{\|v\|, \|v'\|\}$.

Denoting $X := X_1 \times X_2$ and the norm $\|(u,v)\|_X = \max\left\{\|u\|_{X_1}, \|v\|_{X_2}\right\}$, it is clear that $(X, \|.\|_X)$ is a Banach space.

A pair of functions (u,v) is a solution of problem (6.1)-(6.2) if $(u,v) \in X$ and verifies conditions (6.1) and (6.2).

L^1-Carathéodory functions are applied in the sense of Definition 3.1, but adapted now for functions $g : [a,b] \times \mathbb{R}^4 \to \mathbb{R}$ and condition (iii) replaced by:

for each $\rho > 0$, there exists a positive function $\phi_\rho \in L^1([a,b])$ and for $(t,y,z,w) \in \mathbb{R}^4$ such that

$$\max\{|t|, |y|, |z|, |w|\} < \rho, \tag{6.3}$$

one has

$$|g(x,t,y,z,w)| \le \phi_\rho(t), \text{ a.e. } x \in [a,b].$$

Lemma 6.1. *A pair of functions $(u,v) \in X$ is a solution of problem (6.1)-(6.2) if and only if,*

$$u(x) = A_1 + B_1(x-a)$$

$$+ \sum_{x_k < x} [I_{0k}(x_k, u(x_k), u'(x_k)) + I_{1k}(x_k, u(x_k), u'(x_k))(x - x_k)]$$

$$- (x-a) \sum_{k=1}^{n} I_{1k}(x_k, u(x_k), u'(x_k))$$

$$+ \int_a^b G_1(x,s) \, f(s, u(s), u'(s), v(s), v'(s)) \, ds,$$

with $G_1(x,s)$ given by

$$G_1(x,s) = \begin{cases} a - s, & a \le x \le s \le b, \\ a - x, & a \le s \le x \le b, \end{cases} \tag{6.4}$$

and

$$v(x) = A_2 + \frac{B_2 - A_2}{b - a}(x - a)$$

$$+ \sum_{\tau_j < x} [J_{0j}(\tau_j, v(\tau_j), v'(\tau_j)) + J_{1j}(\tau_j, v(\tau_j), v'(\tau_j))(x - \tau_j)]$$

$$- \frac{x-a}{b-a} \sum_{j=1}^{m} [J_{0j}(\tau_j, v(\tau_j), v'(\tau_j)) + J_{1j}(\tau_j, v(\tau_j), v'(\tau_j))(x - \tau_j)]$$

$$+ \int_a^b G_2(x,s) \, h(s, u(s), u'(s), v(s), v'(s)) \, ds,$$

with $G_2(x, s)$ *defined by*

$$G_2(x, s) = \frac{1}{a-b} \begin{cases} (a-s)(b-x) & a \le x \le s \le b, \\ (x-a)(b-s) & a \le s \le x \le b. \end{cases} \tag{6.5}$$

The proof follows standard calculus and it is omitted.

6.2 Main theorem

The main result will provide the existence of, at least, a solution for problem (6.1)-(6.2).

Theorem 6.1. *Let* $f, h : [a, b] \times \mathbb{R}^4 \to \mathbb{R}$ *be* L^1-*Carathéodory functions and* $I_{ik}, J_{ij} : [a, b] \times \mathbb{R}^2 \to \mathbb{R}$ *be continuous functions for* $i = 0, 1$, $k = 1, 2, ..., n$, $j = 1, 2, ..., m$. *Moreover, assume that there is* $R > 0$, *such that*

$$\max \left\{ \begin{array}{c} |A_1| + |B_1|(b-a) + \sum_{k=1}^{n} [\varphi_{0k} + 2(b-a)\varphi_{1k}] \\ + \int_a^b M_1(s)\phi_R(s)ds, \\ \\ |B_1| + 2\sum_{k=1}^{n} \varphi_{1k} + \int_a^b \phi_R(s)ds, \\ \\ |A_2| + |B_2 - A_2| + 2\sum_{\tau_j < x} [\varphi_{0j}^* + \varphi_{1j}^*(b-a)] \\ + \int_a^b M_2(s)\psi_R(s)ds, \\ \\ \frac{|B_2 - A_2|}{b-a} + \frac{1}{b-a}\sum_{j=1}^{m} \varphi_{0j}^* + 3\sum_{j=1}^{m} \varphi_{1j}^* \\ + \frac{1}{b-a}\int_a^b \left|\frac{\partial G_2}{\partial x}(x, s)\right| \psi_R(s)ds \end{array} \right\} < R, \tag{6.6}$$

where M_1, $M_2 \in L^\infty(\mathbb{R})$ *given by*,

$$M_1(s) := \sup_{x \in [a,b]} |G_1(x, s)|, \quad M_2(s) := \sup_{x \in [a,b]} |G_2(x, s)|, \tag{6.7}$$

ϕ_R, $\psi_R \in L^1([a, b])$ *positive functions*, $(t, x, y, z, w) \in \mathbb{R}^5$ *with*

$$|f(t, x, y, z, w)| \le \phi_R(t), \quad a.e. \ x \in [a, b], \tag{6.8}$$

$$|h(t, x, y, z, w)| \le \phi_R(t), \quad a.e. \ x \in [a, b], \tag{6.9}$$

and φ_{ik}, φ_{ij}^* *positive constants such that with*

$$|I_{ik}(x_k, u(x_k), u'(x_k)| \le \varphi_{ik} \ and \ |J_{ij}(\tau_j, v(\tau_j), v'(\tau_j))| \le \varphi_{ij}^*,$$

for $i = 0, 1$, $k = 1, 2, ..., n$, *and* $j = 1, 2, ..., m$.

Then, there is at least a pair of functions $(u, v) \in X$ *that is a solution of (6.1)-(6.2).*

Proof. Define the operators $T_1 : X \to X_1$, $T_2 : X \to X_2$, and $T : X \to X$ by

$$T(u, v) = (T_1(u, v), T_2(u, v)), \tag{6.10}$$

with

$$
\begin{aligned}
(T_1(u, v))(x) &= A_1 + B_1(x - a) \\
&\quad + \sum_{x_k < x} \left[I_{0k}(x_k, u(x_k), u'(x_k)) + I_{1k}(x_k, u(x_k), u'(x_k))(x - x_k) \right] \\
&\quad - (x - a) \sum_{k=1}^{n} I_{1k}(x_k, u(x_k), u'(x_k)) \\
&\quad + \int_a^b G_1(x, s) \, f(s, u(s), u'(s), v(s), v'(s)) \, ds, \\
(T_2(u, v))(x) &= A_2 + \frac{B_2 - A_2}{b - a}(x - a) \\
&\quad + \sum_{\tau_j < x} \left[J_{0j}(\tau_j, v(\tau_j), v'(\tau_j)) + J_{1j}(\tau_j, v(\tau_j), v'(\tau_j))(x - \tau_j) \right] \\
&\quad - \frac{x - a}{b - a} \sum_{j=1}^{m} \left[J_{0j}(\tau_j, v(\tau_j), v'(\tau_j)) + J_{1j}(\tau_j, v(\tau_j), v'(\tau_j))(x - \tau_j) \right] \\
&\quad + \int_a^b G_2(x, s) \, h(s, u(s), u'(s), v(s), v'(s)) \, ds,
\end{aligned}
$$

where $G_1(x, s)$ and $G_2(x, s)$ are given by (6.4) and (6.5), respectively.

By Lemma 6.1, it is obvious that the fixed points of T are solutions of (6.1)-(6.2), so we shall prove that T has a fixed point, following, for clearness, several steps.

Step 1: *T is well defined and continuous in X.*

As $f, h : [a, b] \times \mathbb{R}^4 \to \mathbb{R}$ are L^1-Carathéodory functions, then $T_1(u, v) \in PC_1^1$ and $T_2(u, v) \in PC_2^1$. In fact, $T(u, v) = (T_1(u, v), T_2(u, v))$ is continuous and

$$
\begin{aligned}
(T_1(u, v))'(x) &= B_1 + \sum_{x_k < x} I_{1k}(x_k, u(x_k), u'(x_k)) - \sum_{k=1}^{n} I_{1k}(x_k, u(x_k), u'(x_k)) \\
&\quad - \int_a^x f(s, u(s), u'(s), v(s), v'(s)) \, ds,
\end{aligned}
$$

$$(T_2(u,v))'(x) = \frac{B_2 - A_2}{b - a} + \sum_{\tau_j < x} J_{1j}(\tau_j, v(\tau_j), v'(\tau_j))$$

$$- \frac{1}{b - a} \sum_{j=1}^{m} \left[J_{0j}(\tau_j, v(\tau_j), v'(\tau_j)) + J_{1j}(\tau_j, v(\tau_j), v'(\tau_j))(x - \tau_j) \right]$$

$$- \frac{x - a}{b - a} \sum_{j=1}^{m} J_{1j}(\tau_j, v(\tau_j), v'(\tau_j))$$

$$+ \int_a^b \frac{\partial G_2}{\partial x}(x, s) \, h(s, u(s), u'(s), v(s), v'(s)) \, ds,$$

with

$$\frac{\partial G_2}{\partial x}(x, s) = \frac{1}{b - a} \begin{cases} s - a, & a \le x \le s \le b, \\ b - s, & a \le s \le x \le b. \end{cases} \tag{6.11}$$

Therefore, $T_1(u, v) \in X_1$, $T_2(u, v) \in X_2$ and $T(u, v) \in X$.

Step 2: *TB is uniformly bounded in $B \subset X$.*

Let B be a bounded set of X. Then, there is $\rho_1 > 0$ such that

$$\max \left\{ \|u\|_{X_1}, \|v\|_{X_2} \right\} < \rho_1. \tag{6.12}$$

Moreover,

$$\|T_1(u, v)(x)\|$$

$$\le \sup_{x \in [a,b]} \left(|A_1| + |B_1||x - a| \right.$$

$$+ \sum_{x_k < x} | \left[I_{0k}(x_k, u(x_k), u'(x_k)) + I_{1k}(x_k, u(x_k), u'(x_k))(x - x_k) \right] |$$

$$+ |(x - a)| \sum_{k=1}^{n} |I_{1k}(x_k, u(x_k), u'(x_k))|$$

$$+ \int_a^b |G_1(x, s)| \, |f(s, u(s), u'(s), v(s), v'(s))| \, ds$$

$$\le |A_1| + |B_1|(b - a) + \sum_{k=1}^{n} \left[\varphi_{0k} + 2(b - a)\varphi_{1k} \right]$$

$$+ \int_a^b M_1(s)\phi_{\rho_1}(s)ds < +\infty,$$

$$\| (T_1 (u, v))' (x) \| \leq \sup_{x \in [a,b]} \left(|B_1| + \sum_{x_k < x} |I_{1k}(x_k, u(x_k), u'(x_k))| \right.$$

$$\left. + \sum_{k=1}^{n} |I_{1k}(x_k, u(x_k), u'(x_k))| + \int_a^b |f(s, u(s), u'(s), v(s), v'(s))| \, ds \right)$$

$$\leq |B_1| + 2 \sum_{k=1}^{n} \varphi_{1k} + \int_a^b \phi_{\rho_1}(s) ds < +\infty,$$

$$\| T_2 (u, v) (x) \| \leq \sup_{x \in [a,b]} \left(|A_2| + \frac{|B_2 - A_2|}{b - a} |x - a| \right.$$

$$+ \sum_{\tau_j < x} \left[|J_{0j}(\tau_j, v(\tau_j), v'(\tau_j)) + J_{1j}(\tau_j, v(\tau_j), v'(\tau_j)) (x - \tau_j) | \right]$$

$$+ \frac{|x - a|}{b - a} \sum_{j=1}^{m} |J_{0j}(\tau_j, v(\tau_j), v'(\tau_j)) + J_{1j}(\tau_j, v(\tau_j), v'(\tau_j)) (x - \tau_j) |$$

$$\left. + \int_a^b |G_2(x, s)| \, |h(s, u(s), u'(s), v(s), v'(s))| \, ds \right)$$

$$\leq |A_2| + |B_2 - A_2| + 2 \sum_{\tau_j < x} \left[\varphi_{0j}^* + \varphi_{1j}^* (b - a) \right]$$

$$+ \int_a^b M_2(s) \psi_{\rho_1}(s) ds < +\infty,$$

and, by (6.11),

$$\| (T_2 (u, v))' (x) \| \leq \sup_{x \in [a,b]} \left(\frac{|B_2 - A_2|}{b - a} + \sum_{\tau_j < x} |J_{1j}(\tau_j, v(\tau_j), v'(\tau_j))| \right.$$

$$+ \frac{1}{b - a} \sum_{j=1}^{m} |J_{0j}(\tau_j, v(\tau_j), v'(\tau_j)) + J_{1j}(\tau_j, v(\tau_j), v'(\tau_j)) (x - \tau_j) |$$

$$+ \frac{|x - a|}{b - a} \sum_{j=1}^{m} |J_{1j}(\tau_j, v(\tau_j), v'(\tau_j))|$$

$$\left. + \frac{1}{b - a} \int_a^b \left| \frac{\partial G_2}{\partial x}(x, s) \right| |h(s, u(s), u'(s), v(s), v'(s))| ds \right)$$

$$\leq \frac{|B_2 - A_2|}{b - a} + \frac{1}{b - a} \sum_{j=1}^{m} \varphi_{0j}^* + 3 \sum_{j=1}^{m} \varphi_{1j}^*$$

$$+ \frac{1}{b - a} \int_a^b \left| \frac{\partial G_2}{\partial x}(x, s) \right| \psi_{\rho_1}(s) ds < +\infty.$$

So, TB is uniformly bounded on X.

Step 3: T *is equicontinuous on each interval* $]x_k, x_{k+1}] \times]\tau_j, \tau_{j+1}]$, *that is, $T_1 B$ is equicontinuous on each interval* $]x_k, x_{k+1}]$, *for* $k = 0, 1, ..., n$, *with* $x_0 = a$ *and* $x_{n+1} = b$, *and $T_2 B$ is equicontinuous on each interval* $]\tau_j, \tau_{j+1}]$, *for* $j = 0, 1, ..., m$, *with* $\tau_0 = a$ *and* $\tau_{m+1} = b$.

Consider $J \subseteq]x_k, x_{k+1}]$ and $\iota_1, \iota_2 \in J$ such that $\iota_1 \leq \iota_2$.
So, by the continuity of G_1

$$
\begin{aligned}
&|T_1(u, v)(\iota_1) - T_1(u, v)(\iota_2)| \\
&= \Big| B_1(\iota_1 - \iota_2) \\
&\quad + \sum_{x_k < \iota_1} [I_{0k}(x_k, u(x_k), u'(x_k)) + I_{1k}(x_k, u(x_k), u'(x_k))(\iota_1 - x_k)] \\
&\quad - (\iota_1 - \iota_2) \sum_{k=1}^{n} I_{1k}(x_k, u(x_k), u'(x_k)) \\
&\quad - \sum_{x_k < \iota_2} [I_{0k}(x_k, u(x_k), u'(x_k)) + I_{1k}(x_k, u(x_k), u'(x_k))(\iota_2 - x_k)] \\
&\quad + \int_a^b [G_1(\iota_1, s) - G_1(\iota_2, s)] f(s, u(s), u'(s), v(s), v'(s)) \, ds \Big| \to 0,
\end{aligned}
$$

as $\iota_1 \to \iota_2$,

$$
\begin{aligned}
&\left| (T_1(u, v)(\iota_1))' - (T_1(u, v)(\iota_2))' \right| \\
&= \Big| \sum_{x_k < \iota_1} I_{1k}(x_k, u(x_k), u'(x_k)) - \sum_{x_k < \iota_2} I_{1k}(x_k, u(x_k), u'(x_k)) \\
&\quad - \int_{\iota_1}^{\iota_2} f(s, u(s), u'(s), v(s), v'(s)) \, ds \Big| \to 0,
\end{aligned}
$$

as $\iota_1 \to \iota_2$,

$$
\begin{aligned}
&|T_2(u, v)(\iota_1) - T_2(u, v)(\iota_2)| \\
&= \Big| \frac{B_2 - A_2}{b - a}(\iota_1 - \iota_2) \\
&\quad + \sum_{x_k < \iota_1} [J_{0j}(\tau_j, v(\tau_j), v'(\tau_j)) + J_{1j}(\tau_j, v(\tau_j), v'(\tau_j))(\iota_1 - \tau_j)] \\
&\quad + \frac{\iota_2 - \iota_1}{b - a} \sum_{j=1}^{m} J_{0j}(\tau_j, v(\tau_j), v'(\tau_j)) - \frac{\iota_1 - a}{b - a} \sum_{j=1}^{m} J_{1j}(\tau_j, v(\tau_j), v'(\tau_j))(\iota_1 - \tau_j)
\end{aligned}
$$

$$- \sum_{x_k < \iota_2} \left[J_{0j}(\tau_j, v(\tau_j), v'(\tau_j)) + J_{1j}(\tau_j, v(\tau_j), v'(\tau_j)) (\iota_2 - \tau_j) \right]$$

$$+ \frac{\iota_2 - a}{b - a} \sum_{j=1}^{m} J_{1j}(\tau_j, v(\tau_j), v'(\tau_j)) (\iota_2 - \tau_j)$$

$$+ \left. \int_a^b \left[G_2(\iota_1, s) - G_2(\iota_2, s) \right] h(s, u(s), u'(s), v(s), v'(s) \, ds \right|$$

$\to 0$, as $\iota_1 \to \iota_2$,

and

$$\left| (T_2(u, v)(\iota_1))' - (T_2(u, v)(\iota_2))' \right|$$

$$= \left| \sum_{\tau_j < \iota_1} J_{1j}(\tau_j, v(\tau_j), v'(\tau_j)) - \frac{1}{b-a} \sum_{j=1}^{m} J_{1j}(\tau_j, v(\tau_j), v'(\tau_j)) (\iota_1 - \iota_2) \right.$$

$$+ \frac{(\iota_1 - \iota_2)}{b - a} \sum_{j=1}^{m} J_{1j}(\tau_j, v(\tau_j), v'(\tau_j)) - \sum_{\tau_j < \iota_2} J_{1j}(\tau_j, v(\tau_j), v'(\tau_j))$$

$$\left. + \frac{1}{b-a} \int_{\iota_1}^{\iota_2} \frac{\partial G_2}{\partial x}(x, s) h(s, u(s), u'(s), v(s), v'(s)) \, ds \right| \to 0,$$

as $\iota_1 \to \iota_2$, and $\frac{\partial G_2}{\partial x}$ given by (6.11).

Step 4: TB *is equiconvergent, that is,* $T_1 B$ *is equiconvergent at* $x = x_k^+$ *for* $k = 0, 1, \ldots, n$, *and* $T_2 B$ *is equiconvergent at* $\tau = \tau_j^+$ *for* $\tau = 1, \ldots, m$. In fact,

$$\left| T_1(u, v)(x) - T_1(u, v)(x_k^+) \right|$$

$$= \left| B_1(x - x_k^+) + \sum_{x_k < x} \left[I_{0k}(x_k, u(x_k), u'(x_k)) + I_{1k}(x_k, u(x_k), u'(x_k)) (x - x_k) \right] \right.$$

$$- (x - x_k^+) \sum_{k=1}^{n} I_{1k}(x_k, u(x_k), u'(x_k))$$

$$- \sum_{x_k < x_k^+} \left[I_{0k}(x_k, u(x_k), u'(x_k)) + I_{1k}(x_k, u(x_k), u'(x_k)) (x_k^+ - x_k) \right]$$

$$\left. + \int_a^b \left[G_1(x, s) - G_1(x_k^+, s) \right] f(s, u(s), u'(s), v(s), v'(s)) \, ds \right| \to 0,$$

uniformly as $x \to x_k^+$ and

$$\left| (T_1(u,v)(x))' - (T_1(u,v)(x_k^+))' \right|$$

$$= \left| \sum_{x_k < x} I_{1k}(x_k, u(x_k), u'(x_k)) - \sum_{x_k < x_k^+} I_{1k}(x_k, u(x_k), u'(x_k)) \right.$$

$$\left. - \int_{x_k^+}^x f(s, u(s), u'(s), v(s), v'(s)) \, ds \right| \to 0,$$

uniformly as $x \to x_k^+$. So, $T_1 B$ is equiconvergent at $x = x_{k+}$ for $k = 0, 1, \ldots, n$.

Similarly,

$$\left| T_2(u,v)(\tau) - T_2(u,v)(\tau_j^+) \right|$$

$$= \left| \frac{B_2 - A_2}{b-a}(\tau - \tau_j^+) + \sum_{\tau_j < \tau} \left[J_{0j}(\tau_j, v(\tau_j), v'(\tau_j)) + J_{1j}(\tau_j, v(\tau_j), v'(\tau_j))(\tau - \tau_j) \right] \right.$$

$$+ \frac{\tau_j^+ - \tau}{b-a} \sum_{j=1}^m J_{0j}(\tau_j, v(\tau_j), v'(\tau_j)) - \frac{\tau - a}{b-a} \sum_{j=1}^m J_{1j}(\tau_j, v(\tau_j), v'(\tau_j))(\tau - \tau_j)$$

$$- \sum_{\tau_j < \tau_j^+} \left[J_{0j}(\tau_j, v(\tau_j), v'(\tau_j)) + J_{1j}(\tau_j, v(\tau_j), v'(\tau_j))(\tau_j^+ - \tau_j) \right]$$

$$+ \frac{\tau_j^+ - a}{b-a} \sum_{j=1}^m J_{1j}(\tau_j, v(\tau_j), v'(\tau_j))(\tau_j^+ - \tau_j)$$

$$\left. + \int_a^b \left[G_2(\tau, s) - G_2(\tau_j^+, s) \right] h(s, u(s), u'(s), v(s), v'(s)) \, ds \right| \to 0,$$

uniformly as $\tau \to \tau_j^+$, and

$$\left| (T_2(u,v)(\tau))' - (T_2(u,v)(\tau_j^+))' \right|$$

$$= \left| \sum_{\tau_j < \tau} J_{1j}(\tau_j, v(\tau_j), v'(\tau_j)) - \frac{1}{b-a} \sum_{j=1}^m J_{1j}(\tau_j, v(\tau_j), v'(\tau_j))(\tau - \tau_j^+) \right.$$

$$+ \frac{(\tau - \tau_j^+)}{b-a} \sum_{j=1}^m J_{1j}(\tau_j, v(\tau_j), v'(\tau_j)) - \sum_{\tau_j < \tau_j^+} J_{1j}(\tau_j, v(\tau_j), v'(\tau_j))$$

$$+ \frac{1}{b-a} \left(\int_{\tau_j^+}^\tau (b-s) h(s, u(s), u'(s), v(s), v'(s)) ds \right.$$

$$\left. \left. + \int_{\tau_j^+}^\tau (s-a) h(s, u(s), u'(s), v(s), v'(s)) ds \right) \right| \to 0,$$

uniformly as $\tau \to \tau_j^+$. Then, $T_2 B$ is equiconvergent at $\tau = \tau_j^+$ for $\tau = 1, \ldots, m$.

Therefore, T_1 and T_2 map bounded sets into relatively compact sets, that is, $T_1 : X \to X_1$ and $T_2 : X \to X_2$ are compacts. Therefore, $T : X \to X$ is compact (for details see [173], Lemma 2.4).

Step 5: $T : X \to X$ *has a fixed point.*

In order to apply Schauder's fixed point theorem for operator $T(u, v)$, we need to prove that $TD \subset D$, for some closed, bounded and convex $D \subset X$.

Consider

$$D := \{(u, v) \in X : \|(u, v)\|_X \le \rho_2\},$$

with $\rho_2 > 0$ such that

$$\rho_2 := \max \left\{ \begin{array}{c} \rho_1, \\[2mm] |A_1| + |B_1| (b - a) + \sum_{k=1}^n [\varphi_{0k} + 2(b - a)\varphi_{1k}] \\[2mm] + \int_a^b M_1(s)\phi_{\rho_2}(s)ds, \\[2mm] |B_1| + 2 \sum_{k=1}^n \varphi_{1k} + \int_a^b \phi_{\rho_2}(s)ds, \\[2mm] |A_2| + |B_2 - A_2| + 2 \sum_{\tau_j < x} [\varphi_{0j}^* + \varphi_{1j}^* (b - a)] \\[2mm] + \int_a^b M_2(s)\psi_{\rho_2}(s)ds, \\[2mm] \frac{|B_2 - A_2|}{b - a} + \frac{1}{b - a} \sum_{j=1}^m \varphi_{0j}^* + 3 \sum_{j=1}^m \varphi_{1j}^* \\[2mm] + \frac{1}{b - a} \int_a^b \left| \frac{\partial G_2}{\partial x}(x, s) \right| \psi_{\rho_2}(s)ds \end{array} \right\}, \quad (6.13)$$

with ρ_1 given by (11.4), according to Step 2 and M_1, M_2 are given by (6.7). Following similar arguments as in Step 2, pursue

$$\begin{aligned} \|T(u, v)\|_X &= \|(T_1(u, v), T_2(u, v))\|_X \\ &= \max \left\{ \|T_1(u, v)\|_{X_1}, \|T_2(u, v)\|_{X_2} \right\} \\ &= \max \left\{ \|T_1(u, v)\|, \|(T_1(u, v))'\|, \right. \\ &\qquad \left. \|T_2(u, v)\|, \|(T_2(u, v))'\| \right\} \\ &\le \rho_2, \end{aligned}$$

and $TD \subset D$.

By Schauder's fixed point theorem, the operator T, given by (6.10) has a fixed point (u_0, v_0). Thus, by Lemma 6.1, the problem (6.1)-(6.2) has at least a pair solution $(u, v) \in X$.   □

6.3   Example

Consider the second-order coupled system with the mixed boundary conditions

$$
\begin{cases}
u''(x) = \dfrac{sgn\left(x-\frac{1}{2}\right)u(x)v(x)+(u'(x))^2v'(x)}{100}, \ x \in \left]0,\, 1\right[\\[2ex]
v''(x) = \dfrac{-(x+1)(v'(x))^2u(x)+(u'(x))^3e^{-v(x)}}{500} \\[2ex]
u(0) = 1, \ u'(1) = \frac{1}{2}, \\[1ex]
v(0) = 1, \ v(1) = 2,
\end{cases}
\tag{6.14}
$$

and the generalized impulsive conditions

$$
\begin{cases}
\Delta u(x_k) = \dfrac{(u(x_k))^2(1-x_k)+u'(x_k)}{100}, \ k = 1,2,3, \\[2ex]
\Delta u'(x_k) = \dfrac{\sum_{k=1}^{3} x_k u(x_k)u'(x_k)}{100}, \\[2ex]
\Delta v(\tau_j) = \dfrac{\tau_j|v(\tau_j)|v'(\tau_j)}{500}, \ j = 1,2, \\[2ex]
\Delta v'(\tau_j) = \dfrac{\sum_{j=1}^{2}(1-\tau_j)\frac{v(\tau_j)}{2}(v'(\tau_j))^2}{500},
\end{cases}
\tag{6.15}
$$

keep on $0 < x_1 < x_2 < x_3 < 1$, $0 < \tau_1 < \tau_2 < 1$.

This problem is a particular case of system (6.1)-(6.2) with

$$
f(x,\alpha,\beta,\gamma,\delta) = \frac{sgn\left(x-\frac{1}{2}\right)\alpha\beta+\gamma^2\delta}{100},
$$

$$
h(x,\alpha,\beta,\gamma,\delta) = \frac{-(x+1)\delta^2\alpha+\gamma^3 e^{-\beta}}{500},
$$

$$
A_1 = 1, \ B_1 = \frac{1}{2}, \ A_2 = 1, \ B_2 = 2,
$$

$$
I_{0k}(x_k,\alpha,\beta) = \frac{\alpha^2(1-x_k)+\gamma}{100}, \ I_{1k}(x_k,\alpha,\beta) = \frac{\sum_{k=1}^{3} x_k\alpha\beta}{100},
$$

$$
J_{0j}(\tau_j,\gamma,\delta) = \frac{\tau_j|\beta|\delta}{500}, \ J_{1j}(\tau_j,\gamma,\delta) = \frac{\sum_{j=1}^{2}(1-\tau_j)\frac{\beta}{2}\delta^2}{500}.
$$

In fact, f, h are L^1-Carathéodory functions in $[0, 1]$, with $\rho > 0$ such that

$$\max\left\{|\alpha|, |\beta|, |\gamma|, |\delta|\right\} < \rho,$$

we have

$$|f(x, \alpha, \beta, \gamma, \delta)| \leq \frac{\rho^2 + \rho^3}{100} := \phi_\rho(x),$$

$$|h(x, \alpha, \beta, \gamma, \delta)| \leq \frac{(x+1)\rho^3 + \rho^4}{500} := \psi_\rho(x).$$

Moreover, for $\rho \in [2.0348, 14.805]$, the assumptions of Theorem 6.1 hold and, therefore problem (6.14)-(6.15) has at least a solution $(u, v) \in X$.

6.4   The transverse vibration system of elastically coupled double-string

Consider the transverse vibration system of elastically coupled double-string with damping. The strings have the same length L, are attached by a viscoelastic element and stretched at a constant tension, according to Figure 6.1.

By [213], the system of elastically coupled double-string stationary model is given by the second-order nonlinear coupled system

$$\begin{cases} S_1 u''(x) - K\left(u(x) - v(x)\right) = -l_1(x), \\[2mm] S_2 v''(x) - K\left(v(x) - u(x)\right) = -l_2(x), \end{cases} \tag{6.16}$$

where $x \in [0, L]$,

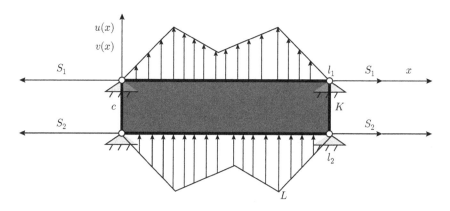

Fig. 6.1   Elastically coupled double-string.

- $u(x)$, $v(x)$ are the transverse deflections of strings u and v, respectively;
- $l_1(x)$ and $l_2(x)$ are the excited distributed loads;
- K is the modulus of Kelvin-Voigt viscoelastic;
- S_1, S_2 are the string tensions of u and v, respectively.

Adding to the system (6.16) the boundary conditions,

$$\begin{cases} u(0) = 0, \ u'(L) = B_1, \\ \\ v(0) = 0, \ v(L) = 0. \end{cases} \tag{6.17}$$

We remark that strings u and v have different behaviors at the end points. Moreover, we consider impulsive conditions that may depend on the string deflections and on the slope of the corresponding deflections,

$$\begin{cases} I_{0k}(x_k, u(x_k), u'(x_k)) = \eta_1 u(x_k) + \eta_2 u'(x_k) + x_k, \\ \\ I_{1k}(x_k, u(x_k), u'(x_k)) = \eta_3 u(x_k) + \eta_4 u'(x_k) + x_k, \\ \\ J_{0j}(\tau_j, v(\tau_j), v'(\tau_j)) = \eta_5 v(\tau_j) + \eta_6 v'(\tau_j) + \tau_j, \\ \\ J_{1j}(\tau_j, v(\tau_j), v'(\tau_j)) = \eta_7 v(\tau_j) + \eta_8 v'(\tau_j) + \tau_j, \end{cases} \tag{6.18}$$

with $B_1, \eta_i \in \mathbb{R}$, $i = 1, .., 8$, $k = 1, .., n$ and $j = 1, .., m$. In this way, we have a particular case of (6.1)-(6.2).

In the system (6.16)-(6.18), it follows that,

$$f(x, \alpha, \beta, \gamma, \delta) = \tfrac{1}{S_1} \left[K \left(\alpha - \gamma \right) - l_1(x) \right],$$

$$h(x, \alpha, \beta, \gamma, \delta) = \tfrac{1}{S_2} \left[+K \left(\gamma - \alpha \right) - l_2(x) \right],$$

and

$$I_{0k}(x_k, \alpha, \gamma) = \eta_1 \alpha + \eta_2 \gamma, \ I_{1k}(x_k, \alpha, \gamma) = \eta_3 \alpha + \eta_4 \gamma + x_k,$$

$$J_{0j}(\tau_j, \beta, \delta) = \eta_5 \beta + \eta_6 \delta, \ J_{1j}(\tau_j, \beta, \delta) = \eta_7 \beta + \eta_8 \delta + \tau_j,$$

with $k = 1, \ldots, n$ and $j = 1, \ldots, m$.

Notice that, f, h are L^1-Carathéodory functions in $[0, L] \times \mathbb{R}^4$, with $\rho > 0$ and

$$\max \left\{ |\alpha|, |\beta|, |\gamma|, |\delta| \right\} < \rho,$$

$$|f(x, \alpha, \beta, \gamma, \delta)| \leq \frac{1}{|S_1|} \left[2|K|\rho + |l_1(x)| \right] := \phi_\rho(x),$$

$$|h(x, \alpha, \beta, \gamma, \delta)| \leq \frac{1}{|S_2|} \left[2|K|\rho + |l_2(x)| \right] := \psi_\rho(x).$$

As the impulsive conditions, I_{ik}, for $i = 1, 2$, $k = 1, .., n$ and J_{ij}, for $j = 1, .., m$, are continuous functions on $[0, L] \times \mathbb{R}^2$. Then, for the values of S_1, S_2, K, $l_1(x)$, $l_2(x)$ and ρ such that the assumptions of Theorem 6.1 are satisfied, the problem (6.16)-(6.18) has at least solution $(u, v) \in X$.

Chapter 7

Impulsive coupled systems on the half-line

Boundary-value problems in unbounded domains, can be applied to a large variety of contexts, (see for instance, [28,81,111,160,172,173,185,216,252]).

Some examples of impulsive effects can be found in [76], where Dishliev et al., make a very complete explanation of these equations, and show also some applications on pharmacokinetic model, logistic model, Gompertz model (mathematical model for a time series), Lotka-Volterra model and population dynamics. In [104], Guo uses the fixed point theory to investigate the existence and uniqueness of solutions of two-point boundary value problems for second-order nonlinear impulsive integro-differential equations on infinite intervals in a Banach space. The same author, in [105], by a comparison result, obtains the existence of maximal and minimal solutions of initial value problems for a class of second-order impulsive integro-differential equations in a Banach space. In [158], Lee and Liu study the existence of extremal solutions for a class of singular boundary value problems of second-order impulsive differential equations. In [195], Minhós, and Carapinha study separated impulsive problems with a fully third-order differential equation, including an increasing homeomorphism, and impulsive conditions given by generalized functions. In [217], Pang et al., consider a second-order impulsive differential equation with integral boundary conditions, where they proposed some sufficient conditions for the existence of solutions, by using the method of upper and lower solutions and Leray-Schauder degree theory. In [157], Lee and Lee combine the method of upper and lower solutions with fixed point index theorems on a cone to study the existence of positive solutions for a singular two point boundary value problem of second-order impulsive equation with fixed moments.

In [237], Shen and Wang investigate the boundary value problem with impulse effect

$$
\begin{aligned}
&x''(t) = f(t, x(t), x'(t)), \quad t \in J, \ t \neq t_k \\
&\Delta x(t_k) = I_k(x(t_k)), \quad k = 1, 2, \ldots, p, \\
&\Delta x'(t_k) = J_k(x(t_k), x'(t_k)), \quad k = 1, 2, \ldots, p, \\
&g(x(0), x'(0)) = 0, \quad h(x(1), x'(1)) = 0,
\end{aligned}
$$

where $J = [0, 1]$, $f : J \times \mathbb{R}^2 \to \mathbb{R}$ is continuous, $I_k \in C(\mathbb{R})$, $J_k \in C(\mathbb{R}^2)$ for $1 \leq k \leq p$, $0 = t_0 < t_1 < \cdots < t_p < t_{p+1} = 1$, $\Delta x(t_k) = x(t_k^+) - x(t_k^-)$, denotes the jump of $x(t)$ at $t = t_k$, $x(t_k^+)$ and $x(t_k^-)$ represent the right and left limits of $x(t)$ at $t = t_k$, respectively, $\Delta x'(t_k) = x'(t_k^+) - x'(t_k^-)$ where

$$
x'(t_k^-) = \lim_{h \to 0^-} h^{-1}[x(t_k + h) - x(t_k)], \quad x'(t_k^+) = \lim_{h \to 0^+} h^{-1}[x(t_k + h) - x(t_k)],
$$

and $g, h : \mathbb{R}^2 \to \mathbb{R}$ are continuous functions.

In [253], Wang, Zhang and Liang, consider the initial value problem for second-order impulsive integro-differential equations, which nonlinearity depend on the first derivative, in a Banach space E:

$$
\begin{cases}
x''(t) = f(t, x(t), x'(t), Tx(t), Sx(t)), \quad t \neq t_k, \ k = 1, 2, \ldots, m, \\
\Delta x(t_k) = I_k(x(t_k), x'(t_k)), \quad k = 1, 2, \ldots, m, \\
\Delta x'(t_k) = \overline{I_k}(x(t_k), x'(t_k)), \quad k = 1, 2, \ldots, m, \\
x(0) = x_0, \quad x'(0) = x_0^*,
\end{cases}
$$

where $f \in C[J \times E^4, E]$, $J = [0, 1]$, $0 < t_0 < t_1 < \cdots < t_k < \cdots < t_m < 1$. $I_k, \overline{I_k} \in C[E^2, E]$, $k = 1, 2, \ldots, m$, $x_0, x_0^* \in E$, θ denotes the zero element of E, $J' = J \setminus \{t_1, t_2, \ldots, t_m\}$ and $J_0 = [0, t_1]$, $J_k = (t_k, t_{k+1}]$, $k = 1, 2, \ldots, m$, $t_{m+1} = 1$,

$$
Tx(t) = \int_0^t k(t, s)x(s)ds, \quad Sx(t) = \int_0^1 h(t, s)x(s)ds, \quad \forall t \in J,
$$

where $k \in C[D, \mathbb{R}_+]$, $D = \{(t, s) : J \times J | t \geq s\}$, $h \in C[J \times J, \mathbb{R}_+]$, $\mathbb{R}_+ = [0, +\infty)$.

In [171], the authors study the existence of multiple and single positive solutions of two-point boundary value problems for the systems of nonlinear second-order singular and impulsive differential equations:

$$
\begin{cases}
-u''(t) = h_1(t)f_1(t, u, v), \quad t \in J', \\
-v''(t) = h_2(t)f_2(t, u, v), \quad t \in J' \\
-\Delta u'|_{t=t_k} = I_{1,k}(u(t_k)), \quad k = 1, 2, \ldots, m, \\
-\Delta v'|_{t=t_k} = I_{2,k}(v(t_k)), \quad k = 1, 2, \ldots, m, \\
\alpha u(0) - \beta u'(0) = 0, \quad \alpha v(0) - \beta v'(0) = 0, \\
\gamma u(1) + \delta u'(1) = 0, \quad \gamma v(1) + \delta v'(1) = 0,
\end{cases}
$$

where $\alpha, \beta, \gamma, \delta \geq 0$, $\rho = \beta\gamma + \alpha\gamma + \alpha\delta > 0$, $J = (0, 1)$, $0 < t_1 < ... < t_m < 1$, $J' = J \setminus \{t_1, t_2, ..., t_m\}$, $\overline{J} = [0, 1]$, $f_i \in C(\overline{J} \times (\mathbb{R}^+)^2, \mathbb{R}^+)$, $I_{i,k} \in C(\mathbb{R}^+, \mathbb{R}^+)$, $h_i \in (J, (0, +\infty))$, $(i = 1, 2)$, and may be singular at $t = 0$ or $t = 1$, $\mathbb{R}^+ = [0, +\infty)$.

In [145], we found the study of second-order nonlinear differential equation

$$(p(t)u'(t))' = f(t, u(t)), \quad t \in (0, \infty) \setminus \{t_1, t_2, ..., t_n\},$$

where $f : [0, +\infty) \times \mathbb{R} \to \mathbb{R}$ is continuous, $p \in [0, +\infty) \cap C(0, +\infty)$ and $p(t) \geq 0$ for all $t > 0$, with the impulsive conditions

$$\Delta u'(t_k) = I_k(u(t_k)), \quad k = 1, ...n,$$

where $I_k : \mathbb{R} \to \mathbb{R}$, $k = 1, ..., n$, are Lipschitz continuous, $n \geq 1$, and the boundary conditions

$$\alpha u(0) - \beta \lim_{t \to 0^+} p(t)u'(t) = 0,$$
$$\gamma \lim_{t \to \infty} u(t) + \delta \lim_{t \to \infty} p(t)u'(t) = 0.$$

In order to ensure that the non-resonant scenario is considered, the condition

$$\rho = \gamma\beta + \alpha\delta + \alpha\gamma \int_0^\infty \frac{d\tau}{p(\tau)} \neq 0$$

is imposed.

In [156], the authors prove the existence of multiple positive solutions for a singular Gelfand type boundary value problem with the following second-order impulsive differential system:

$$u''(t) + \lambda h_1(t)f(u(t), v(t)) = 0, \quad t \in (0, 1), \quad t \neq t_1,$$

$$v''(t) + \mu h_2(t)g(u(t), v(t)) = 0, \quad t \in (0, 1), \quad t \neq t_1,$$

$$\Delta u \mid_{t=t_1} = I_u(u(t_1)), \quad \Delta v \mid_{t=t_1} = I_v(v(t_1)),$$

$$\Delta u' \mid_{t=t_1} = N_u(u(t_1)), \quad \Delta v' \mid_{t=t_1} = N_v(v(t_1)),$$

$$u(0) = a \geq 0, \quad v(0) = b \geq 0, \quad u(1) = c \geq 0, \quad v(1) = d \geq 0,$$

where λ, μ are positive real parameters, $\Delta u \mid_{t=t_1} = u(t_1^+) - u(t_1^-)$, $\Delta u' \mid_{t=t_1} = u'(t_1^+) - u'(t_1^-)$, $f, g \in C(\mathbb{R}^2, (0, \infty))$, $I_u, I_v \in C(\mathbb{R}, \mathbb{R})$ satisfying $I_u(0) = 0 = I_v(0)$, $N_u, N_v \in C(\mathbb{R}, (-\infty, 0])$, and $h_1, h_2 \in C((0, 1), (0, \infty))$.

Inspired by these works, we follow arguments and techniques considered in [194] and [197], in particular, about impulsive problems on the half-line and second-order coupled systems on the half-line, respectively. However, it is the first time where the existence of solutions is obtained for impulsive coupled systems, with generalized jump conditions in half-line and with full nonlinearities, that depend on the unknown functions and their first derivatives.

In particular, in the present chapter, we consider the second-order impulsive coupled system in half-line composed by the differential equations, for $t \in [0, +\infty[$,

$$\begin{cases} u''(t) = f(t, u(t), v(t), u'(t), v'(t)), \ t \neq t_k, \\ v''(t) = h(t, u(t), v(t), u'(t), v'(t)), \ t \neq \tau_j, \end{cases} \tag{7.1}$$

where $f, h : [0, +\infty[\times \mathbb{R}^4 \to \mathbb{R}$ are L^1-Carathéodory functions, the boundary conditions

$$\begin{cases} u(0) = A_1, \ v(0) = A_2, \\ u'(+\infty) = B_1, \ v'(+\infty) = B_2, \end{cases} \tag{7.2}$$

for $A_1, A_2, B_1, B_2 \in \mathbb{R}$ and the generalized impulsive conditions

$$\begin{cases} \Delta u(t_k) = I_{0k}(t_k, u(t_k), u'(t_k)), \\ \Delta u'(t_k) = I_{1k}(t_k, u(t_k), u'(t_k)), \\ \Delta v(\tau_j) = J_{0j}(\tau_j, v(\tau_j), v'(\tau_j)), \\ \Delta v'(\tau_j) = J_{1j}(\tau_j, v(\tau_j), v'(\tau_j)), \end{cases} \tag{7.3}$$

where, $k, j \in \mathbb{N}$,

$$\Delta u^{(i)}(t_k) = u^{(i)}(t_k^+) - u^{(i)}(t_k^-), \ \Delta v^{(i)}(\tau_j) = v^{(i)}(\tau_j^+) - v^{(i)}(\tau_j^-),$$

$I_{ik}, J_{ij} \in C([0, +\infty[\times \mathbb{R}^2, \mathbb{R}), \ i = 0, 1,$ with t_k, τ_j fixed points such that $0 < t_1 < \cdots < t_k < \cdots, 0 < \tau_1 < \cdots < \tau_j < \cdots$ and

$$\lim_{k \to +\infty} t_k = +\infty, \ \lim_{j \to +\infty} \tau_j = +\infty.$$

Some arguments, based on [206], play a key role, such as: Carathéodory functions and sequences, the equiconvergence at each impulsive moment and at infinity, Banach spaces with weighted norms, and Schauder's fixed point theorem, to prove the existence of solutions.

7.1 Definitions and preliminary results

Define $u(t_k^{\pm}) := \lim\limits_{t \to t_k^{\pm}} u(t)$, $v(\tau_j^{\pm}) := \lim\limits_{t \to \tau_j^{\pm}} v(t)$, and consider the set

$$PC_1\left([0, +\infty[\right) = \left\{ \begin{array}{c} u : u \in C([0, +\infty[\setminus \{t_k\}, \mathbb{R}), u(t_k) = u(t_k^-), \\ u(t_k^+) \text{ exists for } k \in \mathbb{N} \end{array} \right\},$$

$$PC_1^n\left([0, +\infty[\right) = \left\{ u : u^{(n)} \in PC_1\left([0, +\infty[\right) \right\}, \, n = 1, 2,$$

$$PC_2\left([0, +\infty[\right) = \left\{ \begin{array}{c} v : v \in C([0, +\infty[\setminus \{\tau_j\}, \mathbb{R}), v(\tau_j) = v(\tau_j^-), \\ v(\tau_j^+) \text{ exists for } j \in \mathbb{N} \end{array} \right\},$$

and $PC_2^n\left([0, +\infty[\right) = \left\{ v : v^{(n)} \in PC_2\left([0, +\infty[\right) \right\}, \, n = 1, 2.$

Denote the space

$$X_1 := \left\{ x : x \in PC_1^1\left([0, +\infty[\right) : \lim\limits_{t \to +\infty} \frac{x(t)}{1+t} \in \mathbb{R}, \lim\limits_{t \to +\infty} x'(t) \in \mathbb{R} \right\},$$

$$X_2 := \left\{ y : y \in PC_2^1\left([0, +\infty[\right) : \lim\limits_{t \to +\infty} \frac{y(t)}{1+t} \in \mathbb{R}, \lim\limits_{t \to +\infty} y'(t) \in \mathbb{R} \right\},$$

and $X := X_1 \times X_2$.

In fact, X_1, X_2 and X are Banach spaces with the norms

$$\|u\|_{X_1} = \max\left\{ \|u\|_0, \|u'\|_1 \right\}, \quad \|v\|_{X_2} = \max\left\{ \|v\|_0, \|v'\|_1 \right\},$$

and

$$\|(u, v)\|_X = \max\left\{ \|u\|_{X_1}, \|v\|_{X_2} \right\},$$

respectively, where

$$\|w\|_0 := \sup_{t \in [0, +\infty[} \frac{|w(t)|}{1+t} \quad \text{and} \quad \|w\|_1 := \sup_{t \in [0, +\infty[} |w(t)|.$$

In this chapter we consider L^1-Carathéodory functions mentioned in Definition 3.1, now adapted as a function $g : [0, +\infty[\times \mathbb{R}^4 \to \mathbb{R}$ and condition *(iii)* adapted by:

for each $\rho > 0$, there exists a positive function $\phi_\rho \in L^1\left([0, +\infty[\right)$ such that, for a.e. $t \in [0, +\infty[$, $(x, y, z, w) \in \mathbb{R}^4$ with

$$\sup_{t \in [0, +\infty[} \left\{ \frac{|x|}{1+t}, \frac{|y|}{1+t}, |z|, |w| \right\} < \rho, \tag{7.4}$$

one has

$$|g(t, x, y, z, w)| \le \phi_\rho(t). \tag{7.5}$$

Definition 7.1. A sequence $(c_n)_{n \in \mathbb{N}} : [0, +\infty[\times \mathbb{R}^2 \to \mathbb{R}$ is a Carathéodory sequence if it verifies

(i) for each $(a, b) \in \mathbb{R}^2$, $(a, b) \longmapsto c_n(t_n, a, b)$ is continuous for all $n \in \mathbb{N}$;

(ii) for each $\rho > 0$, there are nonnegative constants $\chi_{n,\rho} \geq 0$ with $\sum_{n=1}^{+\infty} \chi_{n,\rho} < +\infty$ such that for $|a| < \rho(1+t)$, $t \in [0, +\infty[$ and $|b| < \rho$ we have $|c_n(t, a, b)| \leq \chi_{n,\rho}$, for every $n \in \mathbb{N}$, $t \in [0, +\infty[$.

Lemma 7.1. *Assume that $f, h : [0, +\infty[\times \mathbb{R}^4 \to \mathbb{R}$ are L^1-Carathéodory functions and $I_{0k}, I_{1k}, J_{0j}, J_{1j} : [0, +\infty[\times \mathbb{R}^2 \to \mathbb{R}$ are Carathéodory sequences, for $k, j \in \mathbb{N}$. Then the system (7.1) with conditions (7.2), (7.3), has a solution $(u, v) \in X$ expressed by*

$$u(t) = A_1 + B_1 t$$
$$+ \sum_{0 < t_k < t < +\infty} [I_{0k}(t_k, u(t_k), u'(t_k)) + I_{1k}(t_k, u(t_k), u'(t_k)) (t - t_k)]$$
$$-t \sum_{k=1}^{+\infty} I_{1k}(t_k, u(t_k), u'(t_k)) + \int_0^{+\infty} G(t, s) f(s, u(s), v(s), u'(s), v'(s)) ds,$$

$$v(t) = A_2 + B_2 t$$
$$+ \sum_{0 < \tau_j < t < +\infty} [J_{0j}(\tau_j, v(\tau_j), v'(\tau_j)) + J_{1j}(\tau_j, v(\tau_j), v'(\tau_j)) (t - \tau_j)]$$
$$-t \sum_{j=1}^{+\infty} J_{1j}(\tau_j, v(\tau_j), v'(\tau_j)) + \int_0^{+\infty} G(t, s) h(s, u(s), v(s), u'(s), v'(s)) ds,$$

where

$$G(t, s) = \begin{cases} -t, & 0 \leq t \leq s \leq +\infty, \\ \\ -s, & 0 \leq s \leq t \leq +\infty. \end{cases} \tag{7.6}$$

The proof follows standard techniques and it is omitted.

Definition 7.2. The operator $T : X \to X$ is said to be compact if $T(D)$ is relatively compact, for $D \subseteq X$.

Schauder's fixed point theorem gives the existence solutions (see, Theorem 5.2).

7.2   Existence result

In this section we prove the existence of solution for the problem (7.1)-(7.3).

Theorem 7.1. *Let $f, h : [0, +\infty[\times \mathbb{R}^4 \to \mathbb{R}$ be L^1-Carathéodory functions and $I_{0k}, I_{1k}, J_{0j}, J_{1j} : [0, +\infty[\times \mathbb{R}^2 \to \mathbb{R}$ are Carathéodory sequences such* *that there is $R > 0$ verifying*

$$\max \left\{ \begin{array}{l} K_1 + \sum_{k=1}^{+\infty} \varphi_{k,R} + 2\sum_{k=1}^{+\infty} \psi_{k,R} + \int_0^{+\infty} \Phi_R(s)ds, \\[2ex] K_2 + \sum_{k=1}^{+\infty} \phi_{j,R} + 2\sum_{k=1}^{+\infty} \vartheta_{j,R} + \int_0^{+\infty} \Psi_R(s)ds, \\[2ex] |B_1| + 2\sum_{k=1}^{+\infty} \psi_{k,R} + \int_0^{+\infty} \Phi_R(s)ds, \\[2ex] |B_2| + 2\sum_{k=1}^{+\infty} \vartheta_{j,R} + \int_0^{+\infty} \Psi_R(s)ds \end{array} \right\} < R,$$

where $\varphi_{k,R}$, $\psi_{k,R}$, $\phi_{j,R}$, $\vartheta_{j,R}$ are nonnegative constants such that

$$\sum_{k=1}^{+\infty} \varphi_{k,R} < +\infty, \ \sum_{k=1}^{+\infty} \psi_{k,R} < +\infty, \ \sum_{j=1}^{+\infty} \phi_{j,R} < +\infty, \ \sum_{j=1}^{+\infty} \vartheta_{j,R} < +\infty, \quad (7.7)$$

$$|I_{0k}(t_k, x, y)| \leq \varphi_{k,R}, \ |I_{1k}(t_k, x, y)| \leq \psi_{k,R},$$
$$\text{for } |x| < R(1 + t_k), \ |y| < R, \ k \in \mathbb{N}, \quad (7.8)$$

$$|J_{0j}(\tau_j, x, y)| \leq \phi_{j,R}, \ |J_{1j}(\tau_j, x, y)| \leq \vartheta_{j,R},$$
$$\text{for } |x| < R(1 + \tau_j), \ |y| < R, \ j \in \mathbb{N}, \quad (7.9)$$

and

$$K_i := \sup_{t \in [0,+\infty[} \left(\frac{|A_i| + |B_i t|}{1 + t} \right), \ i = 1, 2. \quad (7.10)$$

Then there is at least a pair $(u, v) \in (PC_1^2([0, +\infty[) \times PC_2^2([0, +\infty[)) \cap X$, solution of (7.1)-(7.3).

Proof. Define the operators $T_1 : X \to X_1$, $T_2 : X \to X_2$, and $T : X \to X$ by

$$T(u, v) = (T_1(u, v), T_2(u, v)), \quad (7.11)$$

with

$$(T_1(u, v))(t) = A_1 + B_1 t$$
$$+ \sum_{0 < t_k < t < +\infty} [I_{0k}(t_k, u(t_k), u'(t_k)) + I_{1k}(t_k, u(t_k), u'(t_k))(t - t_k)]$$
$$- t \sum_{k=1}^{+\infty} I_{1k}(t_k, u(t_k), u'(t_k)) + \int_0^{+\infty} G(t, s) f(s, u(s), v(s), u'(s), v'(s))ds,$$

$$(T_2 (u, v)) (t) = A_2 + B_2 t$$

$$+ \sum_{0 < \tau_j < t < +\infty} [J_{0j}(\tau_j, v(\tau_j), v'(\tau_j)) + J_{1j}(\tau_j, v(\tau_j), v'(\tau_j)) (t - \tau_j)]$$

$$-t \sum_{j=1}^{+\infty} J_{1j}(\tau_j, v(\tau_j), v'(\tau_j)) + \int_0^{+\infty} G(t, s) h(s, u(s), v(s), u'(s), v'(s)) ds,$$

where $G(t, s)$ is defined in (7.6).

The proof will follow several steps which, for clearness, are detailed for operator $T_1 (u, v)$. The technique for operator $T_2 (u, v)$ is similar.

Step 1: T *is well defined and continuous on* X.

Let $(u, v) \in X$. By the Lebesgue dominated convergence theorem, (7.7) and (7.8),

$$\lim_{t \to +\infty} \frac{T_1 (u, v) (t)}{1 + t} = \lim_{t \to +\infty} \left(\frac{A_1 + B_1 t}{1 + t} \right.$$

$$+ \frac{1}{1 + t} \sum_{0 < t_k < t < +\infty} [I_{0k}(t_k, u(t_k), u'(t_k)) + I_{1k}(t_k, u(t_k), u'(t_k)) (t - t_k)]$$

$$\left. - \frac{t}{1 + t} \sum_{k=1}^{+\infty} I_{1k}(t_k, u(t_k), u'(t_k)) \right)$$

$$+ \int_0^{+\infty} \lim_{t \to +\infty} \frac{G(t, s)}{1 + t} f(s, u(s), v(s), u'(s), v'(s)) ds$$

$$\leq B_1 + \sum_{0 < t_k < t < +\infty} I_{1k}(t_k, u(t_k), u'(t_k)) - \sum_{k=1}^{+\infty} I_{1k}(t_k, u(t_k), u'(t_k))$$

$$+ \int_t^{+\infty} |f(s, u(s), v(s), u'(s), v'(s))| ds$$

$$\leq B_1 + 2 \sum_{k=1}^{+\infty} \psi_{k,\rho} + \int_0^{+\infty} \Phi_\rho(s) ds < +\infty,$$

for $\rho > 0$ given by (7.4) and

$$\lim_{t \to +\infty} (T_1(u, v))' (t) = B_1 + \sum_{0 < t_k < t < +\infty} I_{1k}(t_k, u(t_k), u'(t_k))$$

$$- \sum_{k=1}^{+\infty} I_{1k}(t_k, u(t_k), u'(t_k))$$

$$- \lim_{t \to +\infty} \int_t^{+\infty} f(s, u(s), v(s), u'(s), v'(s)) ds$$

$$\leq B_1 + 2 \sum_{k=1}^{+\infty} \psi_{k,\rho} + \int_0^{+\infty} \Phi_\rho(s) ds < +\infty.$$

So, $T_1 X \subset X_1$. Analogously, $T_2 X \subset X_2$. Therefore, T is well defined in X and, as f and h are L^1-Carathéodory functions, by Definition 7.1, T is continuous.

To prove that TD is relatively compact, for $D \subseteq X$ a bounded subset, it is enough to show that:

(i) TD is uniformly bounded, for D a bounded set in X;

(ii) TD is equicontinuous on each interval $]t_k, t_{k+1}] \times]\tau_j, \tau_{j+1}]$, for $k, j = 1, 2, \ldots$;

(iii) TD is equiconvergent at each impulsive point and at infinity.

Step 2: TD *is uniformly bounded, for D a bounded set in* X.

Let $D \subset X$ be a bounded subset. Thus, there is $\rho_1 > 0$ such that, for $(u, v) \in D$,

$$\|(u, v)\|_X = \max \left\{ \|u\|_{X_1}, \|v\|_{X_2} \right\}$$
$$= \max \left\{ \|u\|_0, \|u'\|_1, \|v\|_0, \|v'\|_1 \right\} < \rho_1. \tag{7.12}$$

As, f is a L^1-Carathéodory function, then

$$\|T_1(u, v)\|_0 = \sup_{t \in [0, +\infty[} \frac{|T_1(u, v)(t)|}{1 + t}$$

$$\leq \sup_{t \in [0, +\infty[} \left(\frac{|A_1| + |B_1 t|}{1 + t} \right.$$

$$+ \frac{1}{1 + t} \sum_{0 < t_k < t < +\infty} |I_{0k}(t_k, u(t_k), u'(t_k)) + I_{1k}(t_k, u(t_k), u'(t_k)) (t - t_k)|$$

$$+ \frac{t}{1 + t} \sum_{k=1}^{+\infty} |I_{1k}(t_k, u(t_k), u'(t_k))| \Bigg)$$

$$+ \int_0^{+\infty} \sup_{t \in [0, +\infty[} \frac{|G(t, s)|}{1 + t} |f(s, u(s), v(s), u'(s), v'(s))| ds$$

$$\leq K_1 + \sup_{t\in[0,+\infty[} \left(\frac{1}{1+t} \sum_{0<t_k<t<+\infty} [\varphi_{k,\rho_1} + \psi_{k,\rho_1}(t-t_k)] \right)$$

$$+ \sup_{t\in[0,+\infty[} \left(\frac{t}{1+t} \sum_{k=1}^{+\infty} \psi_{k,\rho_1} \right) + \int_0^{+\infty} \Phi_{\rho_1}(s)ds$$

$$\leq K_1 + \sum_{k=1}^{+\infty} \varphi_{k,\rho_1} + 2\sum_{k=1}^{+\infty} \psi_{k,\rho_1} + \int_0^{+\infty} \Phi_{\rho_1}(s)ds < +\infty, \forall\,(u,v) \in D,$$

and

$$\| (T_1(u,v))' \|_1 = \sup_{t\in[0,+\infty[} | (T_1(u,v))'(t)|$$

$$\leq |B_1| + \sup_{t\in[0,+\infty[} \sum_{0<t_k<t<+\infty} |I_{1k}(t_k,u(t_k),u'(t_k))|$$

$$+ \sum_{k=1}^{+\infty} |I_{1k}(t_k,u(t_k),u'(t_k))|$$

$$+ \sup_{t\in[0,+\infty[} \int_t^{+\infty} |f(s,u(s),v(s),u'(s),v'(s))|ds$$

$$\leq |B_1| + 2\sum_{k=1}^{+\infty} \psi_{k,\rho_1} + \int_0^{+\infty} \Phi_{\rho_1}(s)ds < +\infty.$$

Therefore, $T_1 D$ is bounded and by similar arguments, $T_2 D$ is also bounded. Furthermore, $\|T(u,v)\|_X < +\infty$, that is TD is uniformly bounded on X.

Step 3: TD *is equicontinuous on each interval* $]t_k,t_{k+1}]\times]\tau_j,\tau_{j+1}]$, *that is,* $T_1 D$ *is equicontinuous on each interval* $]t_k,t_{k+1}]$, *for* $k \in \mathbb{N}$, $0 < t_1 < \cdots < t_k < \cdots$, *and* $T_2 D$ *is equicontinuous on each interval* $]\tau_j,\tau_{j+1}]$, *for* $j \in \mathbb{N}$ *and* $0 < \tau_1 < \cdots < \tau_j < \cdots$.

Consider $I \subseteq]t_k,t_{k+1}]$ and $\iota_1,\iota_2 \in I$ such that $\iota_1 \leq \iota_2$. For $(u,v) \in D$, we have

$$\lim_{\iota_1\to\iota_2} \left| \frac{T_1(u,v)(\iota_1)}{1+\iota_1} - \frac{T_1(u,v)(\iota_2)}{1+\iota_2} \right| \leq \lim_{\iota_1\to\iota_2} \left| \frac{A_1+B_1\iota_1}{1+\iota_1} - \frac{A_1+B_1\iota_2}{1+\iota_2} \right|$$

$$+ \left| \frac{1}{1+\iota_1} \sum_{0<t_k<\iota_1} [I_{0k}(t_k,u(t_k),u'(t_k)) + I_{1k}(t_k,u(t_k),u'(t_k))(\iota_1-t_k)] \right.$$

$$-\frac{1}{1+\iota_2} \sum_{0<t_k<\iota_2} \left[I_{0k}(t_k, u(t_k), u'(t_k)) + I_{1k}(t_k, u(t_k), u'(t_k)) (\iota_2 - t_k) \right]$$

$$-\frac{\iota_1}{1+\iota_1} \sum_{k=1}^{+\infty} I_{1k}(t_k, u(t_k), u'(t_k)) + \frac{\iota_2}{1+\iota_2} \sum_{k=1}^{+\infty} I_{1k}(t_k, u(t_k), u'(t_k)) \Bigg|$$

$$+\int_0^{+\infty} \lim_{\iota_1 \to \iota_2} \left| \frac{G(\iota_1, s)}{1+\iota_1} - \frac{G(\iota_2, s)}{1+\iota_2} \right| \left| f(s, u(s), v(s), u'(s), v'(s)) \right| ds = 0,$$

as $\iota_1 \to \iota_2$, and

$$\lim_{\iota_1 \to \iota_2} \left| (T_1(u, v)(\iota_1))' - (T_1(u, v)(\iota_2))' \right|$$

$$\leq \lim_{\iota_1 \to \iota_2} \Bigg| \sum_{0<t_k<\iota_1} I_{1k}(t_k, u(t_k), u'(t_k)) - \sum_{0<t_k<\iota_2} I_{1k}(t_k, u(t_k), u'(t_k))$$

$$-\int_{\iota_1}^{+\infty} f(s, u(s), v(s), u'(s), v'(s)) ds + \int_{\iota_2}^{+\infty} f(s, u(s), v(s), u'(s), v'(s)) ds \Bigg|$$

$$\leq \lim_{\iota_1 \to \iota_2} \sum_{\iota_1<t_k<\iota_2} |I_{1k}(t_k, u(t_k), u'(t_k))| + \int_{\iota_1}^{\iota_2} |f(s, u(s), v(s), u'(s), v'(s))| ds$$

$$\leq \lim_{\iota_1 \to \iota_2} \sum_{\iota_1<t_k<\iota_2} \psi_{k,\rho_1} + \int_{\iota_1}^{\iota_2} \Phi_{\rho_1}(s) ds = 0.$$

Therefore, $T_1 D$ is equicontinuous on X_1. Similarly, we can show that $T_2 D$ is equicontinuous on X_2, too. Thus, TD is equicontinuous on X.

Step 4: *TD is equiconvergent at each impulsive point and at infinity, that is $T_1 D$, is equiconvergent at $t = t_i^+$, $i = 1, 2, \ldots$, and at infinity, and $T_2 D$, is equiconvergent at $\tau = \tau_l^+$, $l = 1, 2, \ldots$, and at infinity.*

First, let us prove the equiconvergence at $t = t_i^+$, for $i = 1, 2, \ldots$. The proof for the equiconvergence at $\tau = \tau_l^+$, for $l = 1, 2, \ldots$, is analogous. Thus, it follows that

$$\left| \frac{T_1(u, v)(t)}{1+t} - \lim_{t \to t_i^+} \frac{T_1(u, v)(t)}{1+t} \right| \leq \left| \frac{A_1 + B_1 t}{1+t} - \frac{A_1 + B_1 t_i}{1+t_i} \right|$$

$$+\left| \frac{1}{1+t} \sum_{0<t_k<t<+\infty} \left[I_{0k}(t_k, u(t_k), u'(t_k)) + I_{1k}(t_k, u(t_k), u'(t_k)) (t - t_k) \right] \right|$$

$$\left| -\frac{1}{1+t_i} \sum_{0<t_k<t_i^+} \left[I_{0k}(t_k, u(t_k), u'(t_k)) + I_{1k}(t_k, u(t_k), u'(t_k))(t_i - t_k) \right] \right.$$

$$+ \left| -\frac{t}{1+t} \sum_{k=1}^{+\infty} I_{1k}(t_k, u(t_k), u'(t_k)) + \frac{t_i}{1+t_i} \sum_{k=1}^{+\infty} I_{1k}(t_k, u(t_k), u'(t_k)) \right|$$

$$+ \int_0^{+\infty} \left| \frac{G(t,s)}{1+t} - \frac{G(t,s)}{1+t_i} \right| \Phi_{\rho_1}(s)ds \to 0,$$

uniformly on $(u,v) \in D$, as $t \to t_i^+$, for $i = 1, 2, \ldots$ and

$$\left| (T_1(u,v)(t))' - \lim_{t \to t_i^+} (T_1(u,v)(t))' \right|$$

$$= \left| \sum_{0<t_k<t<+\infty} I_{1k}(t_k, u(t_k), u'(t_k)) - \sum_{0<t_k<t_i^+} I_{1k}(t_k, u(t_k), u'(t_k)) \right.$$

$$\left. - \int_t^{+\infty} f(s, u(s), v(s), u'(s), v'(s))ds + \int_{t_i}^{+\infty} f(s, u(s), v(s), u'(s), v'(s))ds \right|$$

$$\leq \left| \sum_{0<t_k<t<+\infty} I_{1k}(t_k, u(t_k), u'(t_k)) - \sum_{0<t_k<t_i^+} I_{1k}(t_k, u(t_k), u'(t_k)) \right|$$

$$+ \left| -\int_t^{+\infty} f(s, u(s), v(s), u'(s), v'(s))ds \right.$$

$$\left. + \int_{t_i}^{+\infty} f(s, u(s), v(s), u'(s), v'(s))ds \right|$$

$$\leq \left| \sum_{0<t_k<t<+\infty} I_{1k}(t_k, u(t_k), u'(t_k)) - \sum_{0<t_k<t_i^+} I_{1k}(t_k, u(t_k), u'(t_k)) \right|$$

$$+ \int_{t_i}^{t} \phi_{\rho_1}(s)ds \to 0,$$

uniformly on $(u,v) \in D$, as $t \to t_i^+$, for $i = 1, 2, \ldots$.

Therefore, $T_1 D$ is equiconvergent at each point $t = t_i^+$, for $i = 1, 2, \ldots$. Analogously, it can be proved that $T_2 D$ is equiconvergent at each point $\tau = \tau_l^+$, for $l = 1, 2, \ldots$.

So, TD is equiconvergent at each impulsive point.

To prove the equiconvergence at infinity, for the operator T_1, we have

$$\left| \frac{T_1(u,v)(t)}{1+t} - \lim_{t \to +\infty} \frac{T_1(u,v)(t)}{1+t} \right| \leq \left| \frac{A_1 + B_1 t}{1+t} - B_1 \right|$$

$$+ \left| \frac{1}{1+t} \sum_{0 < t_k < t < +\infty} \left[I_{0k}(t_k, u(t_k), u'(t_k)) + I_{1k}(t_k, u(t_k), u'(t_k)) (t - t_k) \right] \right.$$

$$\left. - \lim_{t \to +\infty} \frac{1}{1+t} \sum_{0 < t_k < t < +\infty} \left[I_{0k}(t_k, u(t_k), u'(t_k)) + I_{1k}(t_k, u(t_k), u'(t_k)) (t - t_k) \right] \right|$$

$$+ \left| -\frac{t}{1+t} \sum_{k=1}^{+\infty} I_{1k}(t_k, u(t_k), u'(t_k)) + \lim_{t \to +\infty} \frac{t}{1+t} \sum_{k=1}^{+\infty} I_{1k}(t_k, u(t_k), u'(t_k)) \right|$$

$$+ \int_0^{+\infty} \left| \frac{G(t,s)}{1+t} - \lim_{t \to +\infty} \frac{G(t,s)}{1+t} \right| |f(s, u(s), v(s), u'(s), v'(s)) ds|$$

$$\leq \left| \frac{A_1 + B_1 t}{1+t} - B_1 \right|$$

$$+ \left| \frac{1}{1+t} \sum_{0 < t_k < t < +\infty} \left[I_{0k}(t_k, u(t_k), u'(t_k)) + I_{1k}(t_k, u(t_k), u'(t_k)) (t - t_k) \right] \right.$$

$$\left. - \sum_{k=1}^{+\infty} I_{1k}(t_k, u(t_k), u'(t_k)) \right|$$

$$+ \left| -\frac{t}{1+t} \sum_{k=1}^{+\infty} I_{1k}(t_k, u(t_k), u'(t_k)) + \sum_{k=1}^{+\infty} I_{1k}(t_k, u(t_k), u'(t_k)) \right|$$

$$+ \int_0^{+\infty} \left| \frac{G(t,s)}{1+t} - \lim_{t \to +\infty} \frac{G(t,s)}{1+t} \right| \Phi_{\rho_1}(s) ds \to 0,$$

uniformly on $(u,v) \in D$, as $t \to +\infty$.

Analogously,

$$\left| (T_1(u,v)(t))' - \lim_{t \to +\infty} (T_1(u,v)(t))' \right|$$

$$= \left| \sum_{0 < t_k < t < +\infty} I_{1k}(t_k, u(t_k), u'(t_k)) - \sum_{k=1}^{+\infty} I_{1k}(t_k, u(t_k), u'(t_k)) \right.$$

$$\left. - \int_t^{+\infty} f(s, u(s), v(s), u'(s), v'(s)) ds \right|$$

$$- \lim_{t \to +\infty} \sum_{0 < t_k < t < +\infty} I_{1k}(t_k, u(t_k), u'(t_k)) + \lim_{t \to +\infty} \sum_{k=1}^{+\infty} I_{1k}(t_k, u(t_k), u'(t_k))$$

$$+ \lim_{t \to +\infty} \int_t^{+\infty} f(s, u(s), v(s), u'(s), v'(s)) ds \bigg|$$

$$\leq \left| \sum_{k=0}^{+\infty} I_{1k}(t_k, u(t_k), u'(t_k)) - \sum_{k=1}^{+\infty} I_{1k}(t_k, u(t_k), u'(t_k)) \right|$$

$$+ \int_t^{+\infty} |f(s, u(s), v(s), u'(s), v'(s))| ds$$

$$\leq \left| \sum_{0 < t_k < t < +\infty} I_{1k}(t_k, u(t_k), u'(t_k)) - \sum_{k=1}^{+\infty} I_{1k}(t_k, u(t_k), u'(t_k)) \right|$$

$$+ \int_t^{+\infty} \Phi_{\rho_1}(s) ds \to 0,$$

uniformly on $(u, v) \in D$, as $t \to +\infty$.

So, $T_1 D$ is equiconvergent at $+\infty$. Following the same arguments, $T_2 D$ is equiconvergent at $+\infty$, too. Therefore, TD is equiconvergent at $+\infty$.

Therefore, TD is relatively compact and, by Definition 7.2, T is compact.

In order to apply Theorem 7.1, we need the next step:

Step 5: $T\Omega \subset \Omega$ *for some* $\Omega \subset X$ *a closed and bounded set.*

Consider

$$\Omega := \{(u, v) \in E : \|(u, v)\|_X \leq \rho_2\},$$

with $\rho_2 > 0$ such that

$$\rho_2 := \max \left\{ \begin{array}{l} \rho_1, \ K_1 + \sum_{k=1}^{+\infty} \varphi_{k,\rho_2} + 2 \sum_{k=1}^{+\infty} \psi_{k,\rho_2} + \int_0^{+\infty} \Phi_{\rho_2}(s) ds, \\[2mm] K_2 + \sum_{k=1}^{+\infty} \phi_{j,\rho_2} + 2 \sum_{k=1}^{+\infty} \vartheta_{j,\rho_2} + \int_0^{+\infty} \Psi_{\rho_2}(s) ds, \\[2mm] |B_1| + 2 \sum_{k=1}^{+\infty} \psi_{k,\rho_2} + \int_0^{+\infty} \Phi_{\rho_2}(s) ds, \\[2mm] |B_2| + 2 \sum_{k=1}^{+\infty} \vartheta_{j,\rho_2} + \int_0^{+\infty} \Psi_{\rho_2}(s) ds, \end{array} \right\}, \tag{7.13}$$

with $\rho_1 > 0$ given by (7.12). According to Step 2 and K_1, K_2 given by (7.10), we have

$$\|T(u, v)\|_X = \|(T_1(u, v), T_2(u, v))\|_X$$
$$= \max \left\{ \|T_1(u, v)\|_{X_1}, \ \|T_2(u, v)\|_{X_2} \right\} \leq \rho_2.$$

So, $T\Omega \subset \Omega$, and by Theorem 5.2, the operator $T(u,v) = (T_1(u,v),$ $T_2(u,v))$, has a fixed point (u,v).

By standard techniques, and Lemma 7.1, it can be shown that this fixed point is a solution of problem (7.1)-(7.3). □

7.3 Motion of a spring pendulum

Consider the motion of the spring pendulum of a mass attached to one end of a spring and the other end attached to the "ceiling". By [212], this motion can be represented by the modified Bessel equations in the system,

$$
\begin{cases}
(t^3 + 1)\, l''(t) = l(t)\theta'(t) - g\cos(\theta(t)) - \frac{k}{m}(l(t) - l_0),\ t \in [0, +\infty[\\[2mm]
(t^3 + 1)\, \theta''(t) = \dfrac{-gl(t)\sin(\theta(t)) - 2l(t)l'(t)\theta'(t)}{l^2(t)},
\end{cases}
$$

$$(7.14)$$

where:

- $l(t)$, l_0 are the length at time t and the natural length of the spring, respectively;
- $\theta(t)$ is the angle between the pendulum and the vertical;
- m, k, g are the mass, the spring constant and gravitational force, respectively;

together with the boundary conditions

$$
\begin{cases}
l(0) = 0,\ \ \theta(0) = 0, \\
l'(+\infty) = B_1,\ \ \theta'(+\infty) = B_2,
\end{cases}
$$

$$(7.15)$$

with $B_1, B_2 \in [0, \pi]$, and the generalized impulsive conditions

$$
\begin{cases}
\Delta l(t_k) = \dfrac{1}{k^3}(\alpha_1 l(t_k) + \alpha_2 l'(t_k)), \\[2mm]
\Delta l'(t_k) = \dfrac{1}{k^4}(\alpha_3 l(t_k) + \alpha_4 l'(t_k)), \\[2mm]
\Delta \theta(\tau_j) = \dfrac{1}{j^5}(\alpha_5 \theta(\tau_j) + \alpha_6 \theta'(\tau_j)), \\[2mm]
\Delta \theta'(\tau_j) = \dfrac{1}{j^3}(\alpha_7 \theta(\tau_j) + \alpha_8 \theta'(\tau_j)),
\end{cases}
$$

$$(7.16)$$

with $\alpha_i \in \mathbb{R}$, $i = 1, 2, \ldots, 8$ and for $k, j \in \mathbb{N}$, $0 < t_1 < \cdots < t_k < \cdots$, $0 < \tau_1 < \cdots \tau_j < \cdots$.

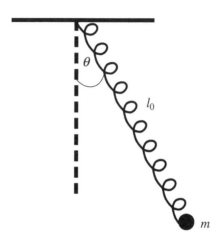

Fig. 7.1 Motion of the spring pendulum.

The system (7.14)-(7.16) is a particular case of the problem (7.1)-(7.3), with

$$f(t, x, y, z, w) = \frac{1}{t^3 + 1}\left(xw - g\cos(y) - \frac{k}{m}(x - l_0)\right),$$

$$h(t, x, y, z, w) = \frac{1}{t^3 + 1}\left(\frac{-gx\sin(y) - 2xzw}{x^2}\right),$$

$$I_{0k}(t_k, x, z) = \frac{1}{k^3}(\alpha_1 x + \alpha_2 z), \quad I_{1k}(t_k, x, z) = \frac{1}{k^4}(\alpha_3 x + \alpha_4 z),$$

$$J_{0j}(\tau_j, y, w) = \frac{1}{j^5}(\alpha_5 y + \alpha_6 w), \quad J_{1j}(\tau_j, y, w) = \frac{1}{j^3}(\alpha_7 y + \alpha_8 w),$$

with $t_k = k$, $\tau_j = j$, $k, j \in \mathbb{N}$, $\alpha_i \in \mathbb{R}$, $i = 1, 2, \ldots, 8$.

Choose $\rho > 0$ such that, for adequate m, k, g, $\phi_\rho(t)$, $\varphi_\rho(t)$, $\alpha_i \in \mathbb{R}$, $i = 1, 2, \ldots, 8$, its holds the following relations

$$f(t, x, y, z, w) \leq \frac{1}{t^3 + 1}\left(\rho^2(1 + t) + g + \frac{k}{m}(\rho(1 + t) + l_0)\right) := \phi_\rho(t),$$

$$h(t, x, y, z, w) \leq \frac{1}{t^3 + 1}\left(\frac{g\rho(1 + t) + 2\rho^3(1 + t)}{l^2(t)}\right)$$

$$\leq \frac{1}{t^3 + 1}\left(\frac{g\rho(1 + t) + 2\rho^3(1 + t)}{l^2(t)}\right)$$

$$\leq \frac{1}{t^3 + 1} \cdot \frac{1}{\left(\min_{t \in [0, +\infty[} l(t)\right)^2}\left(g\rho(1 + t) + 2\rho^3(1 + t)\right)$$

$$:= \varphi_\rho(t),$$

$$I_{0k}(t_k, x, z) \leq \frac{\rho[(\alpha_1(1+k) + \alpha_2)]}{k^3}, \quad I_{1k}(t_k, x, z) \leq \frac{\rho[(\alpha_3(1+k) + \alpha_4)]}{k^4},$$

$$J_{0j}(\tau_j, y, w) \leq \frac{\rho[(\alpha_5(1+k) + \alpha_6)]}{j^5}, \quad J_{1j}(\tau_j, y, w) \leq \frac{\rho[(\alpha_7(1+k) + \alpha_8)]}{j^3}.$$

So, by Theorem 7.1, there is at least a pair $(l, \theta) \in (PC_1^2([0, +\infty[) \times PC_2^2([0, +\infty[)) \cap X$, solution of problem (7.14), (7.15) and (7.16).

Chapter 8

Localization results for impulsive second-order coupled systems on the half-line

Impulsive differential equations are a suitable mathematical tool for modeling the processes and phenomena, which are subjected to some short term external effects during their development. In fact, the duration of these effects is negligible, compared with the total duration of the processes or phenomena. The effects are instantaneous and they take the form of impulses.

Some examples of what has already developed involving impulses, can be seen in [9, 10, 76, 126, 173, 194, 195, 208, 273], and for state impulse problems in [222]. Let us look in more detail at some examples. In [42], we found various theories and techniques on impulsive differential equations and inclusions, as for instance: impulsive functional, neutral and semilinear functional, nonlocal impulsive semilinear differential inclusions, existence results for impulsive functional semilinear, double positive solutions for impulsive boundary value problems and so forth. In [243], the importance of impulsive differential equations, is emphasized with several applications (impulsive biological models, impulsive models in population dynamics, impulsive neural networks, impulsive models in economics).

In [217], Pang and Cai employ the method of upper and lower solutions together with Leray-Schauder degree theory to study the existence of a solution of the impulsive BVP

$$\begin{cases} x''(t) + f(t, x(t), x'(t)) = 0, & t \in J^*, \\ \Delta x(t_k) = I_k(x(t_k)), & k = 1, 2, \ldots, p, \\ \Delta x'(t_k) = J_k(x(t_k), x'(t_k)), & k = 1, 2, \ldots, p, \\ x(0) = x(1) = \int_0^1 g(s)x(s)ds, \end{cases}$$

where $J = [0, 1]$, $J^* = J \setminus \{t_1, t_2, \ldots, t_p\}$, $f : J \times \mathbb{R}^2 \to \mathbb{R}$ is continuous, $I_k, J_k \in C(\mathbb{R})$ for $1 \leq k \leq p$, $g \in L^1[0, 1]$ is nonnegative, $0 = t_0 < t_1 < \cdots < t_p < t_{p+1} = 1$.

Motivated by the works above and applying the method developed in [197, 206], we obtain localization results, via lower and upper solutions for the problem (7.1), (7.2) with two types of impulsive conditions:

First we consider the impulsive conditions given by the generalized functions

$$\begin{cases} \Delta u(t_k) = I_{0k}(t_k, u(t_k)), \\ \Delta v(\tau_j) = J_{0j}(\tau_j, v(\tau_j)) \\ \Delta u'(t_k) = I_{1k}(t_k, u(t_k), u'(t_k)), \\ \Delta v'(\tau_j) = J_{1j}(\tau_j, v(\tau_j), v'(\tau_j)), \end{cases} \tag{8.1}$$

where, $k, j \in \mathbb{N}$,

$$\Delta u^{(i)}(t_k) = u^{(i)}(t_k^+) - u^{(i)}(t_k^-), \quad \Delta v^{(i)}(\tau_j) = v^{(i)}(\tau_j^+) - v^{(i)}(\tau_j^-),$$

for $i = 0, 1$, $I_{0k}, J_{0j} : \mathbb{R}^2 \to \mathbb{R}$ and $I_{1k}, J_{1j} : \mathbb{R}^3 \to \mathbb{R}$ Carathéodory sequences verifying some monotone conditions, with t_k, τ_j fixed points such that $0 = t_0 < t_1 < \cdots < t_k < \cdots$, $0 = \tau_0 < \tau_1 < \cdots < \tau_j < \cdots$ and

$$\lim_{k \to +\infty} t_k = +\infty, \quad \lim_{j \to +\infty} \tau_j = +\infty.$$

Secondly we will have the impulsive effects

$$\begin{cases} \Delta u'(t_k) = I_{0k}^*(t_k, u(t_k), u'(t_k)), \\ \Delta v'(\tau_j) = J_{0j}^*(\tau_j, v(\tau_j), v'(\tau_j)), \\ \Delta u'(t_k) = I_{1k}^*(t_k, u(t_k), u'(t_k)), \\ \Delta v'(\tau_j) = J_{1j}^*(\tau_j, v(\tau_j), v'(\tau_j)), \end{cases} \tag{8.2}$$

with $I_{ik}^*, J_{ij}^* : \mathbb{R}^3 \to \mathbb{R}$, for $i = 0, 1$, as Carathéodory sequences satisfying some growth assumptions.

The method and techniques follow [207].

It should be noted that it is the first time where localization results are considered for coupled systems on the half-line with full nonlinearities, together with generalized infinite impulsive effects.

The main techniques in this chapter make use of Carathéodory functions and sequences, equiconvergence at each impulsive moment and at infinity, following the method suggested in [197], and lower and upper solutions technique combined with Nagumo type condition.

8.1 Preliminary results

In this chapter we adopt the same spaces and norms used in the previous chapter (see Section 7.1). We also consider similar definitions of L^1-Carathéodory functions and Carathéodory sequences (Definition 7.1) as in Chapter 7.

The next two lemmas provide an integral form for the solution of problem (7.1), (7.2), (8.1) or problem (7.1), (7.2), (8.2), as in the previous section.

Lemma 8.1. *Let $f, h : [0, +\infty[\times \mathbb{R}^4 \to \mathbb{R}$ be L^1-Carathéodory functions. Then the system (7.1) with conditions (7.2), (8.1), has a solution $(u(t), v(t))$ expressed by*

$$u(t) = A_1 + B_1 t$$
$$+ \sum_{0 < t_k < t < +\infty} [I_{0k}(t_k, u(t_k)) + I_{1k}(t_k, u(t_k), u'(t_k)) (t - t_k)]$$
$$- t \sum_{k=1}^{+\infty} I_{1k}(t_k, u(t_k), u'(t_k)) + \int_0^{+\infty} G(t, s) f(s, u(s), v(s), u'(s), v'(s)) ds,$$

$$v(t) = A_2 + B_2 t$$
$$+ \sum_{0 < \tau_j < t < +\infty} [J_{0j}(\tau_j, v(\tau_j)) + J_{1j}(\tau_j, v(\tau_j), v'(\tau_j)) (t - \tau_j)]$$
$$- t \sum_{j=1}^{+\infty} J_{1j}(\tau_j, v(\tau_j), v'(\tau_j)) + \int_0^{+\infty} G(t, s) h(s, u(s), v(s), u'(s), v'(s)) ds,$$

where

$$G(t, s) = \begin{cases} -t, & 0 \le t \le s \le +\infty, \\ -s, & 0 \le s \le t \le +\infty. \end{cases} \tag{8.3}$$

Lemma 8.2. *Let $f, h : [0, +\infty[\times \mathbb{R}^4 \to \mathbb{R}$ be L^1-Carathéodory functions. Then the problem (7.1) with conditions (7.2), (8.2), has a solution $(u(t), v(t))$ given by*

$$u(t) = A_1 + B_1 t$$
$$+ \sum_{0 < t_k < t < +\infty} [I_{0k}^*(t_k, u(t_k), u'(t_k)) + I_{1k}^*(t_k, u(t_k), u'(t_k)) (t - t_k)]$$
$$- t \sum_{k=1}^{+\infty} I_{1k}^*(t_k, u(t_k), u'(t_k)) + \int_0^{+\infty} G(t, s) f(s, u(s), v(s), u'(s), v'(s)) ds,$$

$$v(t) = A_2 + B_2 t$$
$$+ \sum_{0 < \tau_j < t < +\infty} \left[J_{0j}^*(\tau_j, v(\tau_j), v'(\tau_j)) + J_{1j}^*(\tau_j, v(\tau_j), v'(\tau_j)) (t - \tau_j) \right]$$
$$-t \sum_{j=1}^{+\infty} J_{1j}^*(\tau_j, v(\tau_j), v'(\tau_j)) + \int_0^{+\infty} G(t,s) h(s, u(s), v(s), u'(s), v'(s)) ds,$$

where $G(t,s)$ is as in (8.3).

8.2　Localization results

The problem (7.1), (7.2), (8.1) is a particular case of the impulsive problem (7.1), (7.2), (7.3) considered in [206], so, in this section, we only prove the localization of solution for problem (7.1), (7.2), (8.1), applying lower and upper solutions method, according to the following definition.

Definition 8.1. A pair of functions $(\alpha_1, \alpha_2) \in (PC_1^2([0, +\infty[) \times PC_2^2([0, +\infty[)) \cap X$ is a lower solution of problem (7.1), (7.2), (8.1) if

$$\alpha_1''(t) \geq f(t, \alpha_1(t), \alpha_2(t), \alpha_1'(t), w), \quad \forall w \in \mathbb{R},$$
$$\alpha_2''(t) \geq h(t, \alpha_1(t), \alpha_2(t), z, \alpha_2'(t)), \quad \forall z \in \mathbb{R},$$
$$\alpha_1(0) \leq A_1, \quad \alpha_2(0) \leq A_2,$$
$$\alpha_1'(+\infty) \leq B_1, \quad \alpha_2'(+\infty) \leq B_2,$$
$$\Delta\alpha_1(t_k) = I_{0k}(t_k, \alpha_1(t_k)),$$
$$\Delta\alpha_2(\tau_j) = J_{0j}(\tau_j, \alpha_2(\tau_j)),$$
$$\Delta\alpha_1'(t_k) > I_{1k}(t_k, \alpha_1(t_k), \alpha_1'(t_k)),$$
$$\Delta\alpha_2'(\tau_j) > J_{1j}(\tau_j, \alpha_2(\tau_j), \alpha_2'(\tau_j)),$$

where $A_1, A_2, B_1, B_2 \in \mathbb{R}$, $k, j \in \mathbb{N}$.

A pair of functions $(\beta_1, \beta_2) \in (PC_1^2([0, +\infty[) \times PC_2^2([0, +\infty[)) \cap X$ is an upper solution of problem (7.1), (7.2), (8.1) if it verifies the reverse inequalities.

Consider the following assumptions

(A1) f, h verify the Nagumo conditions in S_1 and S_2, respectively, with

$$f(t, x, \alpha_2(t), z, w) \geq f(t, x, y, z, w) \geq f(t, x, \beta_2(t), z, w). \quad (8.4)$$

for fixed $(t, x, z, w) \in [0, +\infty[\times \mathbb{R}^3$, and

$$h(t, \alpha_1(t), y, z, w) \geq h(t, x, y, z, w) \geq h(t, \beta_1(t), y, z, w) \qquad (8.5)$$

for fixed $(t, y, z, w) \in [0, +\infty[\times \mathbb{R}^3$;

(A2) For $i = 1, 2$,

$$N_i^* = \max \left\{ |B_i|, \|\alpha_i'\|_1, \|\beta_i'\|_1 \right\}, \qquad (8.6)$$

$I_{1k}(t_k, x, y)$ and $J_{1j}(\tau_j, x, w)$ are nondecreasing on $y \in [-N_1^*, N_1^*]$ and on $w \in [-N_2^*, N_2^*]$, for all $k, j \in \mathbb{N}$, and fixed $x \in \mathbb{R}$.

The main theorem is an existence and localization result:

Theorem 8.1. *Consider $A_1, A_2, B_1, B_2 \in \mathbb{R}$. Let $f, h : [0, +\infty[\times \mathbb{R}^4 \to \mathbb{R}$ be L^1-Carathéodory functions verifying the assumptions of Theorem 7.1, (A1) and (A2), respectively.*

Suppose that $I_{0k}, J_{0j} : \mathbb{R}^2 \to \mathbb{R}$ and $I_{1k}, J_{1j} : \mathbb{R}^3 \to \mathbb{R}$ are Carathéodory sequences, for $k, j \in \mathbb{N}$, verifying the assumptions of Theorem 7.1, (A1) and (A2), respectively.

Assume that there are (α_1, α_2) and (β_1, β_2) coupled lower and upper solutions of problem (7.1), (7.2), (8.1), respectively, such that

$$\alpha_1(t) \leq \beta_1(t), \quad \alpha_2(t) \leq \beta_2(t), \quad t \in [0, +\infty[. \qquad (8.7)$$

Then there is at least a pair $(u(t), v(t)) \in (PC_1^2([0, +\infty[) \times PC_2^2([0, +\infty[)) \cap X$ solution of (7.1), (7.2), (8.1), such that

$$\alpha_1(t) \leq u(t) \leq \beta_1(t), \quad \alpha_2(t) \leq v(t) \leq \beta_2(t), \quad \forall t \in [0, +\infty[. \qquad (8.8)$$

Proof. The existence solution for problem (7.1), (7.2), (8.1), is guaranteed by Theorem 7.1.

To prove the localization part, consider the auxiliary functions $\delta_i : [0, +\infty[\times \mathbb{R} \to \mathbb{R}$, given by

$$\delta_i(t, w) = \begin{cases} \beta_i(t), & w > \beta_i(t) \\ w, & \alpha_i(t) \leq w \leq \beta_i(t) \\ \alpha_i(t), & w < \alpha_i(t), \end{cases}$$

for $i = 1, 2$. The truncated and perturbed auxiliary coupled system is composed by

$$
\begin{cases}
u''(t) = f(t, \delta_1(t, u(t)), \delta_2(t, v(t)), u'(t), v'(t)) \\
\qquad + \dfrac{1}{1+t} \dfrac{u(t) - \delta_1(t, u(t))}{|u(t) - \delta_1(t, u(t))| + 1}, \; t \neq t_k, \\
v''(t) = h(t, \delta_1(t, u(t)), \delta_2(t, v(t)), u'(t), v'(t)) \\
\qquad + \dfrac{1}{1+t} \dfrac{v(t) - \delta_2(t, v(t))}{|v(t) - \delta_2(t, v(t))| + 1}, \; t \neq \tau_j,
\end{cases}
\tag{8.9}
$$

together with conditions (7.2) and the truncated impulsive conditions

$$
\begin{cases}
\Delta u(t_k) = I_{0k}(t_k, \delta_1(t_k, u(t_k))) \\
\Delta v(\tau_j) = J_{0j}(\tau_j, \delta_2(\tau_j, v(\tau_j))), \\
\Delta u'(t_k) = I_{1k}(t_k, \delta_1(t_k, u(t_k)), u'(t_k)) \\
\Delta v'(\tau_j) = J_{1j}(\tau_j, \delta_2(\tau_j, v(\tau_j)), v'(\tau_j)).
\end{cases}
\tag{8.10}
$$

Let $(u(t), v(t))$ be a solution of problem (8.9), (7.2), (8.10). Suppose, by contradiction, that there is $t \in [0, +\infty[$, such that $\alpha_1(t) > u(t)$ and define

$$
\inf_{t \in [0, +\infty[} (u(t) - \alpha_1(t)) := u(t_0) - \alpha_1(t_0) < 0.
$$

Then, $t_0 \neq 0$ and $t_0 \neq +\infty$, by Definition 8.1 and (7.2),

$$
u(0) - \alpha_1(0) = A_1 - \alpha_1(0) \geq 0,
$$

and

$$
u'(+\infty) - \alpha_1'(+\infty) = B_1 - \alpha_1'(+\infty) \geq 0.
$$

Therefore, $t_0 \in]0, +\infty[$, and next two cases can happen:

Case 1: Suppose that there is $p \in \{0, 1, 2, ...\}$, such that $t_0 \in]t_p, t_{p+1}[$. Then,

$$
u'(t_0) = \alpha_1'(t_0), \quad u''(t_0) - \alpha_1''(t_0) \geq 0, \tag{8.11}
$$

and we deduce the following contradiction, by (8.4) and Definition 8.1

$$
\begin{aligned}
0 &\leq u''(t_0) - \alpha_1''(t_0) \\
&= f(t_0, \delta_1(t_0, u(t_0)), \delta_2(t_0, v(t_0)), u'(t_0), v'(t_0)) \\
&\quad + \frac{1}{1+t_0} \frac{u(t_0) - \delta_1(t_0, u(t_0))}{|u(t_0) - \delta_1(t_0, u(t_0))| + 1} - \alpha_1''(t_0) \\
&= f(t_0, \alpha_1(t_0), \delta_2(t_0, v(t_0)), \alpha_1'(t_0), v'(t_0)) \\
&\quad + \frac{1}{1+t_0} \frac{u(t_0) - \alpha_1(t_0)}{|u(t_0) - \alpha_1(t_0)| + 1} - \alpha_1''(t_0)
\end{aligned}
$$

$$\leq f(t_0, \alpha_1(t_0), \alpha_2(t_0), \alpha_1'(t_0), v'(t_0))$$

$$+ \frac{1}{1+t_0} \frac{u(t_0) - \alpha_1(t_0)}{|u(t_0) - \alpha_1(t_0)| + 1} - \alpha_1''(t_0)$$

$$< f(t_0, \alpha_1(t_0), \alpha_2(t_0), \alpha_1'(t_0), v'(t_0)) - \alpha_1''(t_0) \leq 0.$$

Case 2: Assume that there is $p \in \{1, 2, ...\}$ such that

$$\min_{t \in [0, +\infty[} (u(t) - \alpha_1(t)) := u(t_p) - \alpha_1(t_p) < 0. \qquad (8.12)$$

Then by (8.12), we have

$$u'(t_p) \leq \alpha_1'(t_p). \qquad (8.13)$$

As,

$$\Delta(u - \alpha_1)(t_p) = \Delta u(t_p) - \Delta \alpha_1(t_p)$$

$$= I_{0p}(t_p, \delta_1(t_p, u(t_p))) - I_{0p}(t_p, \alpha_1(t_p))$$

$$= I_{0p}(t_p, \alpha_1(t_p)) - I_{0p}(t_p, \alpha_1(t_p)) = 0,$$

by (8.6), (8.10) and (8.13) we have the following contradiction with (8.12),

$$u'(t_p^+) - \alpha_1'(t_p^+) < I_{1p}(t_p, \delta_1(t_p, u(t_p)), u'(t_p)) + u'(t_p)$$

$$- I_{1p}(t_p, \alpha_1(t_p), \alpha_1'(t_p)) - \alpha_1'(t_p)$$

$$= I_{1p}(t_p, \alpha_1(t_p), u'(t_p)) - I_{1p}(t_p, \alpha_1(t_p), \alpha_1'(t_p)) \leq 0.$$

So, $\alpha_1(t) \leq u(t)$, $\forall t \in [0, +\infty[$, and the remaining inequality $u(t) \leq \beta_1(t)$, $\forall t \in [0, +\infty[$, can be proved by same technique.

Applying the method above, it can be shown that $\alpha_2(t) \leq v(t) \leq \beta_2(t)$, $\forall t \in [0, +\infty[$, and, therefore, the problems (8.9), (7.2), (8.10) and (7.1), (7.2), (8.1) are equivalent. $\qquad \square$

8.3 Localization result for more general impulsive conditions

The solvability of problem (7.1), (7.2), (8.2) under the adequate assumptions is guaranteed by Theorem 7.1. To show the localization part, we define lower and upper solutions in a more general way than Definition 8.1, as it follows:

Definition 8.2. A pair of functions $(\alpha_1, \alpha_2) \in (PC_1^2([0, +\infty[) \times PC_2^2([0, +\infty[)) \cap X$ is a lower solution of problem (7.1), (7.2), (8.2) if

$$\alpha_1''(t) \geq f(t, \alpha_1(t), \alpha_2(t), \alpha_1'(t), \alpha_2'(t)),$$

$$\alpha_2''(t) \geq h(t, \alpha_1(t), \alpha_2(t), \alpha_1'(t), \alpha_2'(t)),$$

$$\alpha_1(0) \leq A_1, \quad \alpha_2(0) \leq A_2,$$
$$\alpha_1'(+\infty) \leq B_1, \quad \alpha_2'(+\infty) \leq B_2,$$
$$\Delta\alpha_1(t_k) \leq I_{0k}^*(t_k, \alpha_1(t_k), \alpha_1'(t_k)),$$
$$\Delta\alpha_2(\tau_j) \leq J_{0j}^*(\tau_j, \alpha_2(\tau_j), \alpha_2'(\tau)),$$
$$\Delta\alpha_1'(t_k) > I_{1k}^*(t_k, \alpha_1(t_k), \alpha_1'(t_k)),$$
$$\Delta\alpha_2'(\tau_j) > J_{1j}^*(\tau_j, \alpha_2(\tau_j), \alpha_2'(\tau)),$$

where $A_1, A_2, B_1, B_2 \in \mathbb{R}$.

A pair of functions $(\beta_1, \beta_2) \in (PC_1^2([0, +\infty[) \times PC_2^2([0, +\infty[)) \cap X$ is an upper solution of problem (7.1), (7.2), (8.2) if it verifies the reverse inequalities.

Theorem 8.2. *Consider $A_i, B_i \in \mathbb{R}$, for $i = 1, 2$. Let $f, h : [0, +\infty[\times \mathbb{R}^4 \to \mathbb{R}$ be L^1-Carathéodory functions verifying the assumptions of Theorem 7.1 and $I_{ik}^*, J_{ij}^* : \mathbb{R}^3 \to \mathbb{R}$ be Carathéodory sequences, for $i = 0, 1$.*

Assume that:

- *there are (α_1, α_2) and (β_1, β_2) coupled lower and upper solutions of problem (7.1), (7.2), (8.2), respectively, such that*

$$\alpha_1'(t) \leq \beta_1'(t), \quad \alpha_2'(t) \leq \beta_2'(t), \ t \in [0, +\infty[; \tag{8.14}$$

- *f and h verify*

$$f(t, \alpha_1(t), \alpha_2(t), z, \alpha_2'(t)) \geq f(t, x, y, z, w)$$
$$\geq f(t, \beta_1(t), \beta_2(t), z, \beta_2'(t)), \tag{8.15}$$

for fixed $(t, z) \in [0, +\infty[\times \mathbb{R}$ and

$$h(t, \alpha_1(t), \alpha_2(t), \alpha_1'(t), w) \geq h(t, x, y, z, w)$$
$$\geq h(t, \beta_1(t), \beta_2(t), \beta_1'(t), w), \tag{8.16}$$

for fixed $(t, w) \in [0, +\infty[\times \mathbb{R}$;
- *the impulsive conditions satisfy*

$$I_{0k}^*(t_k, \alpha_1(t_k), \alpha_1'(t_k)) \leq I_{0k}^*(t_k, x, z) \leq I_{0k}^*(t_k, \beta_1(t_k), \beta_1'(t_k)), \tag{8.17}$$

for $\alpha_1(t_k) \leq x \leq \beta_1(t_k)$, $\alpha_1'(t_k) \leq z \leq \beta_1'(t_k)$ and $k \in \mathbb{N}$,

$$J_{0j}^*(\tau_j, \alpha_2(\tau_j), \alpha_2'(\tau_j)) \leq J_{0j}^*(\tau_j, y, w) \leq J_{0j}^*(\tau_j, \beta_2(\tau_j), \beta_2'(\tau_j)), \tag{8.18}$$

for $\alpha_2(\tau_j) \leq y \leq \beta_2(\tau_j)$, $\alpha_2'(\tau_j) \leq w \leq \beta_2'(\tau_j)$ and $\tau_j \in \mathbb{N}$,

$$I_{1k}^*(t_k, \alpha_1(t_k), z) \geq I_{1k}^*(t_k, x, z) \geq I_{1k}^*(t_k, \beta_1(t_k), z), \tag{8.19}$$

for $\alpha_1(t) \leq x \leq \beta_1(t)$, $k \in \mathbb{N}$ and fixed $z \in \mathbb{R}$,

$$J_{1j}^*(\tau_j, \alpha_2(\tau_j), w) \geq J_{1j}^*(\tau_j, y, w) \geq J_j^*(\tau_j, \beta_2(\tau_j), w), \qquad (8.20)$$

for $\alpha_2(t) \leq y \leq \beta_2(t)$, $j \in \mathbb{N}$ and fixed $w \in \mathbb{R}$.

Then there is at least a pair $(u(t), v(t)) \in (PC_1^2([0, +\infty[) \times PC_2^2([0, +\infty[)) \cap X$ solution of (7.1), (7.2), (8.2), such that

$$\alpha_1^{(i)}(t) \leq u^{(i)}(t) \leq \beta_1^{(i)}(t), \quad \alpha_2^{(i)}(t) \leq v^{(i)}(t) \leq \beta_2^{(i)}(t), \quad i = 0, 1, \forall t\, [0, +\infty[. \tag{8.21}$$

Proof. To prove the localization part given by (8.21), consider the auxiliary functions $\delta_i^j : [0, +\infty[\times \mathbb{R} \to \mathbb{R}$, given by

$$\delta_i^j(t, w) = \begin{cases} \beta_i^{(j)}(t), & w > \beta_i^{(j)}(t) \\ w, & \alpha_i^{(j)}(t) \leq w \leq \beta_i^{(j)}(t) \\ \alpha_i^{(j)}(t), & w < \alpha_i^{(j)}(t), \end{cases}$$

for $i = 1, 2$, $j = 0, 1$, and the truncated and perturbed coupled system

$$\begin{cases} u''(t) = f(t, \delta_1^0(t, u(t)), \delta_2^0(t, v(t)), \delta_1^1(t, u'(t)), \delta_2^1(t, v'(t))) \\ \qquad + \frac{1}{1+t} \frac{u'(t) - \delta_1^1(t, u'(t))}{|u'(t) - \delta_1^1(t, u'(t))| + 1} \\ \\ v''(t) = h(t, \delta_1^0(t, u(t)), \delta_2^0(t, v(t)), \delta_1^1(t, u'(t)), \delta_2^1(t, v'(t))) \\ \qquad + \frac{1}{1+t} \frac{v'(t) - \delta_2^1(t, v'(t))}{|v'(t) - \delta_2^1(t, v'(t))| + 1}, \end{cases} \tag{8.22}$$

with conditions (7.2) and the truncated impulsive conditions

$$\begin{cases} \Delta u(t_k) = I_{0k}^*(t_k, \delta_1^0(t_k, u(t_k)), u'(t_k)), \\ \Delta v(\tau_j) = J_{0j}^*(\tau_j, \delta_2^0(\tau_j, v(\tau_j)), v'(\tau_j)), \\ \Delta u'(t_k) = I_{1k}^*(t_k, \delta_1^0(t_k, u(t_k)), \delta_1^1(t_k, u'(t_k))), \\ \Delta v'(\tau_j) = J_{1j}^*(\tau_j, \delta_2^0(\tau_j, v(\tau_j)), \delta_2^1(\tau_j, v'(\tau_j))). \end{cases} \tag{8.23}$$

Let $(u(t), v(t))$ be a solution of problem (8.22), (7.2), (8.23).

Suppose, by contradiction, that there is $t \in [0, +\infty[$, such that $\alpha_1'(t) > u'(t)$ and define

$$\inf_{t \in [0, +\infty[} (u'(t) - \alpha_1'(t)) := u'(t_0) - \alpha_1'(t_0) < 0. \tag{8.24}$$

By (7.2) and Definition 8.2, $t_0 \neq +\infty$, as

$$u'(+\infty) - \alpha_1'(+\infty) = B_1 - \alpha_1'(+\infty) \geq 0.$$

If $t_0 = 0$, then $u''(0) - \alpha_1''(0) \geq 0$ and the following contradiction holds, by Definition 8.2, (8.15) and (8.22),

$$
\begin{aligned}
0 &\leq u''(0) - \alpha_1''(0) \\
&= f(0, \delta_1^0(0, u(0)), \delta_2^0(0, v(0)), \delta_1^1(0, u'(0)), \delta_2^1(0, v'(0))) \\
&\quad + \frac{u'(0) - \delta_1^1(0, u'(0))}{|u'(0) - \delta_1^1(0, u'(0))| + 1} - \alpha_1''(0) \\
&\leq f(0, \delta_1^0(0, u(0)), \delta_2^0(0, v(0)), \alpha_1'(t_0), \delta_2^1(0, v'(0))) \\
&\quad + \frac{u'(0) - \alpha_1'(0)}{|u'(0) - -\alpha_1'(0)| + 1} - \alpha_1''(0) \\
&< f(0, \delta_1^0(0, u(0)), \delta_2^0(0, v(0)), \alpha_1'(0), \delta_2^1(0, v'(0))) - \alpha_1''(0) \\
&\leq f(0, \alpha_1(0), \alpha_2(0), \alpha_1'(0), \alpha_2'(0)) - \alpha_1''(0) \leq 0.
\end{aligned}
$$

Therefore, $t_0 \in]0, +\infty[$ and we have three possible cases:

- Assume that there is $p \in \{0, 1, 2, \ldots\}$, such that $t_0 \in]t_p, t_{p+1}[$. Then,

$$
u''(t_0) = \alpha_1''(t_0),
$$

and by the arguments above, it follows a similar contradiction.
- If there is $p \in \{1, 2, \ldots\}$, such that

$$
\min_{t \in [0, +\infty[} (u'(t) - \alpha_1'(t)) := u'(t_p) - \alpha_1'(t_p) < 0. \tag{8.25}
$$

So, by Definition 8.2, (8.19) and (8.23), the following contradiction holds,

$$
\begin{aligned}
0 &\leq \Delta(u - \alpha_1)'(t_p) = \Delta u'(t_p) - \Delta \alpha_1'(t_p) \\
&< I_{1p}^*(t_p, \delta_1^0(t_p, u(t_p)), \delta_1^1(t_p, u'(t_p))) - I_{1p}^*(t_p, \alpha_1(t_p), \alpha_1'(t_p)) \\
&= I_{1p}^*(t_p, \delta_1^0(t_p, u(t_p)), \alpha_1'(t_p)) - I_{1p}^*(t_p, \alpha_1(t_p), \alpha_1'(t_p)) \\
&= I_{1p}^*(t_p, \alpha_1(t_p), \alpha_1'(t_p)) - I_{1p}^*(t_p, \alpha_1(t_p), \alpha_1'(t_p)) \leq 0.
\end{aligned}
$$

- Assume that there is $p \in \{1, 2, \ldots\}$, such that

$$
\inf_{t \in [0, +\infty[} (u'(t) - \alpha_1'(t)) := u'(t_p^+) - \alpha_1'(t_p^+) < 0.
$$

So, there is $\varepsilon > 0$ sufficiently small such that

$$
u'(t) - \alpha_1'(t) < 0, \ u''(t) - \alpha_1''(t) \geq 0, \ \forall t \in]t_p, t_p + \varepsilon[.
$$

So, for $t \in \,]t_p, t_p + \varepsilon[$, we have the contradiction, given by,

$$
\begin{aligned}
0 &\leq u''(t) - \alpha_1''(t) \\
&= f(t, \delta_1^0(t, u(t)), \delta_2^0(t, v(t)), \delta_1^1(t, u'(t)), \delta_2^1(t, v'(t))) \\
&\quad + \frac{u'(t) - \delta_1^1(t, u'(t))}{|u'(t) - \delta_1^1(t, u'(t))| + 1} - \alpha_1''(t) \\
&= f(t, \delta_1^0(t, u(t)), \delta_2^0(t, v(t)), \alpha_1'(t), \delta_2^1(t, v'(t))) \\
&\quad + \frac{u'(t) - \alpha_1'(t)}{|u'(t) - \alpha_1'(t)| + 1} - \alpha_1''(t) \\
&< f(t, \delta_1^0(t, u(t)), \delta_2^0(t, v(t)), \alpha_1'(t), \delta_2^1(t, v'(t))) - \alpha_1''(t) \\
&\leq f(t, \alpha_1(t), \alpha_2(t), \alpha_1'(t), \alpha_2'(t)) - \alpha_1''(t) \leq 0.
\end{aligned}
$$

Therefore, $\alpha_1'(t) \leq u'(t)$, $\forall t \in [0, +\infty[$. Analogously it can be proved that $u(t)' \leq \beta_1'(t)$, $\forall t \in [0, +\infty[$. By the analogous method, we can show that $\alpha_2'(t) \leq v'(t) \leq \beta_2'(t)$, $\forall t \in [0, +\infty[$.

From the integration of the inequality

$$
\alpha_1'(t) \leq u'(t) \leq \beta_1'(t), \; \forall t \in [0, +\infty[,
$$

for $t \in [0, t_1[$, we have, by (7.2) and Definition 8.2,

$$
\alpha_1(t) \leq u(t) + \alpha_1(0) - u(0) = u(t) - \alpha_1(0) - A_1 \leq u(t).
$$

So,

$$
\alpha_1(t) \leq u(t), \; \forall t \in [0, t_1]. \tag{8.26}
$$

Repeating the above process, for $t \in [t_1, t_2[$, it follows that, by (8.23), (8.17), Definition (8.2) and (8.26),

$$
\begin{aligned}
\alpha_1(t) &\leq u(t) + \alpha_1(t_1^+) - u(t_1^+) \\
&= u(t) + \alpha_1(t_1^+) - I_{01}^*(t_1, \delta_1^0(t_1, u(t_1)), u'(t_1)) - u(t_1^-) \\
&\leq u(t) + I_{01}^*(t_1, \alpha_1(t_1), \alpha_1'(t_1)) - I_{01}^*(t_1, \delta_1(t_1, u(t_1)), u'(t_1)) \\
&\quad - u(t_1^-) + \alpha_1(t_1^-) \\
&\leq u(t) - \left(u(t_1^-) - \alpha_1(t_1^-) \right) \leq u(t).
\end{aligned}
$$

So, $\alpha_1(t) \leq u(t)$, $\forall t \in \,]t_1, t_2[$. By iteration, we obtain that $\alpha_1(t) \leq u(t)$, $\forall t \in [0, +\infty[$.

Similarly, it can be shown that, $u(t) \leq \beta_1(t)$, $\forall t \in [0, +\infty[$. Applying the arguments above, we can prove that $\alpha_2(t) \leq v(t) \leq \beta_2(t)$, $\forall t \in [0, +\infty[$. $\square$

8.4 Logging timber by helicopter

Helicopter logging is a system for the removal of trees in cables attached to a helicopter, of felled and bucked logs from areas where some, or all, of the trees have been felled (see [258]).

In [227], the authors compare the impact of helicopter and rubber-tired skidder extraction of timber after harvesting on the structure and function of a blackwater forested wetland. In [138], Jones *et al.*, studied the removal of logs via helicopters advocated to minimize soil damage and facilitate rapid revegetation. Moreover, they also tested the impacts of the helicopter compared to skidder harvesting systems in regeneration, community structure of woody plants and biomass growth in three blackwater streams of floodplains in southern Alabama.

Helicopter logging is considered practical for harvesting high value timber from inaccessible sites and is a preferred alternative, when harvesting timber from environmentally timber salvage sites, or inaccessible sites [256].

In [101], the authors cite countless advantages of logging by helicopter, among which: is best suited for fast removals of timber, especially in places that require it, to reduce fire risk, to limit the spread of pests, where these are prioritized against cost of the operations and is practice in the valorization of timber of these inaccessible places. So, it is an economically viable alternative compared to other forms of timber extraction.

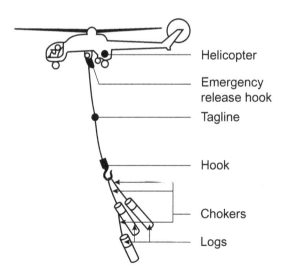

Fig. 8.1 Logging timber by helicopter.

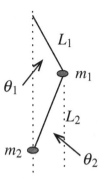

Fig. 8.2   Representation of the cable logging timber by helicopter.

Motivated by these works, we consider a logging timber by helicopter model (see [109]), represented by a second-order linear system

$$\begin{cases} (m_1 + m_2)L_1^2\theta_1''(t) + m_2L_1L_2\theta_2''(t) + (m_1 + m_2)L_1g\theta_1(t) = 0, \\ \\ m_2L_1L_2\theta_1''(t) + m_2L_2^2\theta_2''(t) + m_2L_2g\theta_2(t) = 0, \end{cases}$$

where:

- θ_1 and θ_2 denote the angles of oscillation of the two connecting cables, measured from the gravity vector direction;
- g is the gravitation constant;
- m_1, m_2 denote the masses of the two trees;
- L_1, L_2 are the cable lengths.

The trees are hung in the helicopter in cables and the load for two trees approaching a double pendulum, which oscillates during flight.

From the above linear system, we can obtain a nonlinear system valid for a long-time helicopter flight under forcing terms, given by the second-order nonlinear coupled system on the half-line

$$\begin{cases} \theta_1''(t) = \frac{g}{L_1m_1(t^3+1)}\left[+(m_1+m_2)\theta_1(t) - m_2\theta_2(t)\right] \\ \qquad + g_1(t,\theta_1(t),\theta_2(t),\theta_1'(t),\theta_2'(t)), \\ \\ \theta_2''(t) = \frac{g(m_1+m_2)}{L_2m_1(t^3+1)}(-\theta_1(t)+\theta_2(t)) + g_2(t,\theta_1(t),\theta_2(t),\theta_1'(t),\theta_2'(t)), \end{cases}$$

$$(8.27)$$

for $t \in [0, +\infty[$, $g_1, g_2 : [0, +\infty[\times \mathbb{R}^4 \to \mathbb{R}$ are L^1-Carathéodory forcing functions, to be defined forward, such that $g_1(t, x, y, z, w)$ is nonincreasing in y and $g_2(t, x, y, z, w)$ is nonincreasing in x, together with the boundary conditions

$$\begin{cases} \theta_1(0) = 0, \ \theta_2(0) = 0, \\ \theta_1'(+\infty) = 0, \ \theta_2'(+\infty) = 0. \end{cases} \tag{8.28}$$

Moreover, we consider the impulsive conditions

$$\begin{cases} \Delta\theta_1(k) = \frac{3k^2 + 3k + 1}{k^3(k+1)^3}; \quad \Delta\theta_2(j) = \frac{3j^2 + 3j + 1}{j^3(j+1)^3}, \\ \\ \Delta\theta_1'(k) = \frac{1}{k^3}(-\alpha_1(k) + \alpha_1'(k)); \quad \Delta\theta_2'(j) = \frac{1}{k^j}(-\alpha_2(j) + \alpha_2'(j)), \end{cases} \tag{8.29}$$

for $k, j \in \mathbb{N}$, $0 < t_1 < \cdots < t_k < \cdots$, $0 < \tau_1 < \cdots \tau_j < \cdots$.

The system (8.27)-(8.29) is a particular case of the problem (7.1), (7.2), (8.1), with

$$f(t, x, y, z, w) = \frac{g}{L_1 m_1(t^3 + 1)}[(m_1 + m_2)x - m_2 y] + g_1(t, x, y, z, w),$$

$$h(t, x, y, z, w) = \frac{g(m_1 + m_2)}{L_2 m_1(t^3 + 1)}(-x + y) + g_2(t, x, y, z, w),$$

$$I_{0k}(k, x) = \frac{3k^2 + 3k + 1}{k^3(k+1)^3}, \quad J_{0j}(j, y) = \frac{3j^2 + 3j + 1}{j^3(j+1)^3},$$

$$I_{1k}(k, x, z) = \frac{1}{k^3}(-x + z), \quad J_{1j}(j, y, w) = \frac{1}{j^3}(-y + w),$$

with $k, j \in \mathbb{N}$, $A_1 = A_2 = B_1 = B_2 = 0$. Choose $\rho > 0$ such that, for adequate m_1, m_2, L_1, L_2, g_1 and g_2, the following relations are satisfied

$$|f(t, x, y, z, w)| \leq \frac{g}{L_1 m_1(t^3 + 1)}(m_1 + m_2)\rho(1 + t) + m_2\rho + \Phi_{1_\rho}(t)$$
$$:= \Psi_{1\rho}(t),$$

$$|h(t, x, y, z, w)| \leq \frac{g(m_1 + m_2)}{L_2 m_1(t^3 + 1)}2\rho(1 + t) + \Phi_{2_\rho}(t)$$
$$:= \Psi_{2\rho}(t),$$

where $|g_1(t, x, y, z, w)| \leq \Phi_{1_\rho}(t)$, $|g_2(t, x, y, z, w)| \leq \Phi_{2_\rho}(t)$, for some $\rho > 0$, such that

$$\sup_{t \in [0, +\infty[} \left\{ \frac{|x|}{1 + t}, \frac{|y|}{1 + t}, |z|, |w| \right\} < \rho,$$

and Φ_{i_ρ}, $\Psi_{i\rho}$, $i = 1,2$, are positive functions such that $\Phi_{i_\rho}, \Psi_{i\rho} \in L^1([0,+\infty[)$, and

$$I_{1k}(k,x,z) \le \frac{1}{k^3}\rho(2+k); \quad J_{1j}(j,y,w) \le \frac{1}{j^3}\rho(2+j).$$

The functions $\alpha_i : [0,+\infty[\to \mathbb{R}$, $i = 1,2$, given by

$$\alpha_i(t) = \begin{cases} -1, & t \in [0,1] \\ -\frac{1}{(k+1)^3}, & t \in]k, k+1], \ k > 1, \end{cases}$$

and $\beta_i : [0,+\infty[\to \mathbb{R}$, given by

$$\beta_i(t) = \begin{cases} t, & t \in [0,1] \\ \frac{1}{(k+1)^3}, & t \in]k, k+1], \ k > 1, \end{cases}$$

are, respectively, lower and upper solutions of problem (8.27)-(8.29), satisfying (8.7), assuming that, for $i = 1,2$,

$$g_i(t,-1,-1,0,0) \le 0, \quad g_i(t,1,1,0,0) \ge 0, \quad , \forall t \in [0,1],$$

$$g_i\left(t, -\frac{1}{(k+1)^3}, -\frac{1}{(k+1)^3}, 0, 0\right) \le 0, \forall t \in]k, k+1], \ k > 1,$$

and

$$g_i\left(t, \frac{1}{(k+1)^3}, \frac{1}{(k+1)^3}, 0, 0\right) \ge 0, \forall t \in]k, k+1], \ k > 1.$$

In fact, for the lower solution $(\alpha_1(t), \alpha_2(t))$ and $t \in [0,1]$, we have

$$0 \ge \frac{-gm_1}{L_1 m_1(t^3+1)} + g_1(t,-1,-1,0,0)$$
$$0 \ge g_2(t,-1,-1,0,0),$$

and for $t \in]k, k+1]$, $k > 1$,

$$0 \ge \frac{-gm_1}{L_1 m_1(t^3+1)(k+1)^3} + g_1\left(t, -\frac{1}{(k+1)^3}, -\frac{1}{(k+1)^3}, 0, 0\right)$$
$$0 \ge \frac{-2g(m_1+m_2)}{L_2 m_1(t^3+1)(k+1)^3} + g_2\left(t, -\frac{1}{(k+1)^3}, -\frac{1}{(k+1)^3}, 0, 0\right).$$

For the upper solution $(\beta_1(t), \beta_2(t))$ and $t \in [0,1]$,

$$0 \le \frac{gm_1}{L_1 m_1(t^3+1)} + g_1(t,1,1,0,0)$$
$$0 \le g_2(t,1,1,0,0),$$

and for $t \in \,]k, k+1]$, $k > 1$,

$$0 \leq \frac{gm_1}{L_1 m_1 (t^3 + 1)(k+1)^3} + g_2 \left(t, \frac{1}{(k+1)^3}, \frac{1}{(k+1)^3}, 0, 0 \right)$$

$$0 \leq g_2 \left(t, \frac{1}{(k+1)^3}, \frac{1}{(k+1)^3}, 0, 0 \right).$$

Moreover, remark that, for $i = 1, 2$,

$$\lim_{t \to +\infty} \alpha_i'(t) = \lim_{k \to +\infty} \alpha_i'(t) = 0 \text{ and } \lim_{t \to +\infty} \beta_i'(t) = \lim_{k \to +\infty} \beta_i'(t) = 0.$$

So, by Theorem 8.1, there is at least a pair $(\theta_1, \theta_2) \in (PC_1^2\,([0, +\infty[) \times PC_2^2\,([0, +\infty[)) \cap X$, solution of problem (8.27)-(8.29) and, moreover,

$$-1 \leq \theta_1(t) \leq 1, \quad -1 \leq \theta_2(t) \leq 1, \ \forall t \in [0, 1],$$

$$-\frac{1}{(k+1)^3} \leq \theta_1(t) \leq \frac{1}{(k+1)^3}, \text{ for } t \in \,]k, k+1], \ k = 2, 3, \dots$$

and

$$-\frac{1}{(j+1)^3} \leq \theta_2(t) \leq \frac{1}{(j+1)^3}, \text{ for } j \in \,]j, j+1], \ j = 2, 3, \dots$$

Conclusions and open issues

The third part adapts the methods and techniques developed in the previous ones to impulsive coupled systems with fully differential equations. In fact, it considers several types of boundary conditions and impulsive effects that are defined via very general functions with dependence on the unknown functions and their first derivatives, with bounded or unbounded domains and a finite or infinite number of impulsive moments.

Two main features seem to be relevant for the existence of a solution for impulsive problems:

Equiconvergence at each impulsive moment to control the jumps;

If the impulsive moments are infinite, they must be given as Carathéodory sequences.

The direction fields for improvement given for boundary value problems in previous sections remain valid for impulsive problems. We underline here some open possible tasks characteristic of the impulsive functions:

- As the impulsive functions are evaluated on discrete moments, how to define them to weaken their regularity?
- In the localization technique (see Theorem 8.2), is it possible to have different monotonies on the impulsive functions? And on the nonlinearities?

PART IV

Coupled systems of integral equations

.

Introduction

Integral equations can be seen as generalizations of nonlinear boundary value problems requiring less regularity on the nonlinearities. They often arise in engineering, physical, chemical, inverse and biological problems [17, 112, 244].

Integral equation theory has been an active field of research for many years and its applications have great relevance in mathematics. It was originally from the theory of Fourier[a] integrals. Studying the curve along which a heavy, sliding frictionless particle descends to its lowest position, the Norwegian mathematician Abel[b] was the first to produce an integral equation, in connection with the famous tautochrone problem. In fact, an integral equation of the following form is obtained

$$\int_0^x \frac{\varphi(y)dy}{\sqrt{x-y}} = f(x),$$

where $f(x)$ is the given function and $\varphi(y)$ is the unknown function (see [1, 220, 221, 225]).

Later on, in the end of the nineteenth century, mathematicians Volterra[c] and Fredholm[d] contribute, respectively, to the study of integral equations, presenting a new method to solve the Dirichlet problem.

The main purpose of this Part is to develop in detail the existence theory and to provide some techniques for the systems of integral equations. Concretely we study problems composed of nonlinear higher-order coupled systems of integral equations with complete nonlinearities and the boundary conditions can be expressed at finite or infinite intervals.

---

[a] Jean-Baptiste Joseph Fourier (1768–1830), was a French mathematician and physicist.
[b] Niels Henrik Abel (1802–1829), was a Norwegian mathematician who made pioneering contributions in a variety of fields.
[c] Vito Volterra (1860–1940), was an Italian mathematician and physicist.
[d] Erik Ivar Fredholm (1866–1927), was a Swedish mathematician.

The research methodology followed in this work is based essentially in [209, 240, 241] and:

- Arguments based on *Guo–Krasnosel'skiĭ fixed-point theorem* of *cones theory* for higher-order coupled systems of integral equations;
- *Schauder's fixed point theorem* and the concept of *equiconvergence* at ∞, to recover the compactness of the associated operators.

This fourth Part consists of three chapters which cover the study of integral equations, more concretely existence of solutions of generalized coupled systems of integral equations of Hammerstein type, on bounded and unbounded intervals:

- In the *ninth chapter* we consider the existence of solutions to generalized nonlinear higher-order coupled systems of integral equations of Hammerstein type on the bounded interval [0, 1];
- *The tenth chapter* deals with the existence of solutions for nonlinear higher-order coupled systems of integral equations of Hammerstein-type with sign-changing kernels on [0, 1];
- Lastly, in the *eleventh chapter,* we apply Schauder's fixed point theorem to prove the existence of solutions of generalized coupled systems of integral equations of Hammerstein on the real line.

Chapter 9

Coupled systems of Hammerstein integral equations

In this chapter we deal with generalized coupled systems of integral equations of Hammerstein-type

$$\begin{cases} u_1(t) = \int_0^1 k_1(t,s) \, g_1(s) \, f_1(s, u_1(s), \dots, u_1^{(m_1)}(s), u_2(s), \dots, u_2^{(n_1)}(s)) \, ds, \\ u_2(t) = \int_0^1 k_2(t,s) \, g_2(s) \, f_2(s, u_1(s), \dots, u_1^{(m_2)}(s), u_2(s), \dots, u_2^{(n_2)}(s)) \, ds \end{cases}$$

$$(9.1)$$

where $k_\iota : [0,1]^2 \to \mathbb{R}$, $\iota = 1,2$, are the kernel functions such that $k_\iota \in W^{r_\iota,1}([0,1]^2)$, $r_\iota = \max\{m_\iota, n_\iota\}$, with $m_\iota, n_\iota \geq 0$ positive integers, $g_\iota \in L^1([0,1])$ with $g_\iota(t) \geq 0$ for a.e. $t \in [0,1]$, and $f_\iota : [0,1] \times \mathbb{R}^{m_\iota + n_\iota + 2} \to [0,\infty)$ are L^∞−Carathéodory functions.

The theory of integral equations has been and continues to be a field of many research and applications. These equations are especially relevant in physics and are often used to reformulate or rewrite mathematical problems.

Hammerstein integral equations are special subclasses of nonlinear integral equations of Fredholm-type

$$u(x) = g(x) + \int_a^b k(x, y, u(x), u(y)) f(x, u(y)) dy, \quad x \in [a, b]$$

and their study was initiated by Hammerstein (see [27, 110]).

Hammerstein family of integral equations appears in some mathematical models such as: electrostatic drift waves and low-frequency electromagnetic perturbation (see [88]) and signal theory (see [154]). Other applications and very successful results can be found on [4,66,232] and the references therein.

The existence, uniqueness, multiplicity, positivity and location of solutions are the most studied and predominant elements about Hammerstein integral equations. Citing just a few examples in the literature, we have: [265], where the authors use fixed point index theory to establish their main result, based on *a priori* estimates achieved by nonnegative matrices; in [67], Coclite studies the existence of a positive measurable solution of the Hammerstein equation of the first kind with a singular nonlinear term at the origin; in [53], the authors contribute, by monotone iterative methods, combined with the classical fixed point index, proving two results concerning non-decreasing and non-increasing operators in a shell, in the presence of an upper or a lower solution; in [58], Cardinali *et al.*, examine multivalued Hammerstein integral equations defined in a separable reflexive Banach space, obtaining existence results for convex and nonconvex problems; in [87], the researchers study solutions of the nonlinear Hammerstein integral equation with sign-changing kernels by using a variational principle of Ricceri and critical point theory techniques (they combine the effects of sublinear and superlinear nonlinear terms to establish new existence and multiplicity results); in [272], the authors study the existence and the uniqueness of iterative positive solutions for a class of nonlinear singular integral equations in which the nonlinear terms may be singular in both time and space variables. By using the fixed point theorem of mixed monotone operators in cones, they establish the conditions for the existence and uniqueness of positive solutions to the problem.

In addition, several discretization and numerical methods were also considered on integral equations (see, for instance, [3, 13, 33, 37, 45, 235, 268]).

More specifically, about Hammerstein-type coupled systems of integral equations, we refer to [72], where, by a special cone and using fixed point index theory, Cui and Sun, investigate the existence of positive solutions of singular superlinear coupled integral boundary value problems for differential systems

$$\begin{cases} -x''(t) = f_1(t, x(t), y(t)) \\ -y''(t) = f_2(t, x(t), y(t)), t \in (0, 1), \\ \quad x(0) = y(0) = 0, \\ \quad x(1) = \alpha\,[y]\,, \\ \quad y(1) = \beta\,[x]\,, \end{cases} \tag{9.2}$$

where $f_1, f_2 : (0, 1) \times [0, +\infty)^2 \to [0, +\infty)$ are continuous and may be singular at $t = 0, 1$, and $\alpha[x]$, $\beta[x]$ are bounded linear functionals on $C[0, 1]$ given by

$$\alpha\,[y] = \int_0^1 y(t)dA(t), \ \beta\,[x] = \int_0^1 x(t)dB(t),$$

involving Stieltjes integrals, and A, B are functions of bounded variation with positive measures. Remark that, (9.2) can be reformulated by

$$\begin{cases} x(t) = \int_0^1 G_1(t,s)u(s)ds + \int_0^1 H_1(t,s)v(s)ds, \\ y(t) = \int_0^1 G_2(t,s)v(s)ds + \int_0^1 H_2(t,s)u(s)ds, \end{cases}$$

where

$$G_1(t,s) = \frac{\alpha[t]\,t}{\kappa} \int_0^1 K(s,\tau)dB(\tau), \; ; H_1(t,s) = \frac{t}{\kappa} \int_0^1 K(s,\tau)dA(\tau),$$

$$G_2(t,s) = \frac{\beta[t]\,t}{\kappa} \int_0^1 K(s,\tau)dA(\tau), \; ; H_2(t,s) = \frac{t}{\kappa} \int_0^1 K(s,\tau)dB(\tau),$$

$$K(t,s) = \begin{cases} t(1-s) \; 0 \le t \le s \le 1, \\ s(1-t) \; 0 \le s \le t \le 1. \end{cases}$$

In [259], the authors study the existence and multiplicity of positive solutions for the system of nonlinear Hammerstein integral equations

$$u(x) = \int_0^1 k_1(x,y)\, f_1(y,u(y),v(y),w(y))\, dy,$$

$$v(x) = \int_0^1 k_2(x,y)\, f_2(y,u(y),v(y),w(y))\, dy,$$

$$w(x) = \int_0^1 k_2(x,y)\, f_3(y,u(y),v(y),w(y))\, dy$$

where $k_i \in C([0,1] \times [0,1], \mathbb{R}_+)$ and $f_i \in C([0,1] \times \mathbb{R}_+^3, \mathbb{R}_+)$, for $i = 1,2,3$. The authors use concave functions to characterize the growth and the behaviors of nonlinearities f_1, f_2, f_3, considering three cases: assuming firstly, that all are superlinear; secondly, with all sublinear and the last case with two superlinear and the other one sublinear. Based on *a priori* estimates achieved by Jensen's integral inequality for concave functions, the authors use the fixed point index theory to establish the main result.

Recently, in [131], Infante and Minhós, extending the results on [200], to prove the existence, multiplicity, non-existence and localization results for nontrivial solutions of the system

$$\begin{cases} u(t) = \int_0^1 k_1(t,s)\, g_1(s)\, f_1(s,u(s),u'(s),v(s),v'(s))\, ds, \\ v(t) = \int_0^1 k_2(t,s)\, g_2(s)\, f_2(s,u(s),u'(s),v(s),v'(s))\, ds. \end{cases}$$

To obtain their results, it is assumed some adequate assumptions in order to apply a fixed index theorem and cone theory.

Motivated by these works we consider the coupled integral system (9.1). Our results are based on [200] and [100], extending the results to systems of coupled Hammerstein-type integral equations with nonlinearities in both unknown functions and their first derivatives.

As far as we know, it is the first time where coupled systems contain integral equations with nonlinearities depending on several derivatives of both variables and, moreover, the derivatives can be of different order on each variable and each equation. That is, both equations and both variables can have different regularity. This detail is very important as it allows the application of our results, for example, to boundary value problems with coupled systems composed by differential equations of different orders and distinct boundary conditions on each unknown function. This issue will open new fields of applications to phenomena modelled by coupled systems requiring different types of regularity on each variable.

The arguments in this chapter apply Guo–Krasnosel'skiĭ compression-expansion theory on cones. Moreover, the kernel functions and the corresponding derivatives associated to the integral equations are nonnegative and verify some adequate sign and growth assumptions. The dependence of the derivatives is overcome by the construction of suitable cones taking into account certain conditions of sublinearity/superlinearity at the origin and at $+\infty$.

9.1   Backgrounds and assumptions

In this book we consider the cones defined, for $\iota = 1, 2$, by

$$
K_\iota := \left\{ w \in C^{r_\iota}[0, 1] : \min_{t \in [a_{i\iota}, b_{i\iota}]} w^{(i)}(t) \geq c_{i\iota} \|w^{(i)}\|_{C^{r_\iota}}, \text{ for } i = 0, 1, \ldots, r_\iota \right\},
$$
$$\tag{9.3}$$

where $0 < c_{i\iota} < 1$ and $r_\iota = \max\{m_\iota, n_\iota\}$. The Banach space $C^k[0, 1]$, equipped with the norm $\|\cdot\|_{C^k}$, defined by

$$
\|w\|_{C^k} := \max\left\{ \|w^{(j)}\| : j = 0, 1, \ldots, k \right\}
$$

and $\|y\| := \max_{t \in [0,1]} |y(t)|$.

Moreover, the set $E := K_1 \times K_2$ with the norm

$$
\|(u_1, u_2)\|_E := \max\left\{ \|u_1\|_{C^{r_1}}, \|u_2\|_{C^{r_2}} \right\},
$$
$$\tag{9.4}$$

is a Banach space.

Definition 9.1. A function $h : [0, 1] \times \mathbb{R}^q \to [0, \infty)$, for q a positive integer, is L^∞–Carathéodory if

(i) $h(\cdot, y)$ is measurable for each fixed $y \in \mathbb{R}^q$;

(ii) $h(t, \cdot)$ is continuous for a.e. $t \in [0, 1]$;

(iii) for each $\rho > 0$, there exists a function $\varphi_\rho \in L^\infty([0, 1])$ such that, $h(t, y) \leq \varphi_\rho(t)$ for $y \in [-\rho, \rho]$ and a.e. $t \in [0, 1]$.

The existence tool will be given by Guo-Krasnoselskii results in the expansive and compressive cones theory, and can be easily obtained from Lemma 1.5 (see chapter 1).

In this chapter we will assume the following conditions.

(A1) For $\iota = 1, 2$, the function $k_\iota : [0, 1]^2 \to \mathbb{R}$, $k_\iota \in W^{r_\iota, 1}([0, 1]^2)$, verify for all $\tau \in [0, 1]$,

$$\lim_{t \to \tau} |k_\iota(t, s) - k_\iota(\tau, s)| = 0, \quad \text{for a.e. } s \in [0, 1],$$

and

$$\lim_{t \to \tau} \left| \frac{\partial^i k_\iota}{\partial t^i}(t, s) - \frac{\partial^i k_\iota}{\partial t^i}(\tau, s) \right| = 0, \quad \text{for a.e. } s \in [0, 1] \text{ and } i = 1, \ldots, r_\iota.$$

(A2) For $\iota = 1, 2$, and every $j_\iota = 0, 1, \ldots, r_\iota$, there exist subintervals $[a_{\iota j}, b_{\iota j}] \subseteq [0, 1]$, positive functions $\phi_{\iota j} \in L^\infty[0, 1]$, and constants $c_{\iota j} \in (0, 1]$, such that:

$$0 \leq k_\iota(t, s) \leq \phi_{\iota 0}(s) \text{ for } t \in [0, 1] \text{ and a.e. } s \in [0, 1];$$

$$0 \leq \frac{\partial^i k_\iota}{\partial t^i}(t, s) \leq \phi_{\iota i}(s) \text{ for } t \in [0, 1], \text{ a.e. } s \in [0, 1] \text{ and}$$
$$i = 1, \ldots, r_\iota;$$

$$k_\iota(t, s) \geq c_{\iota 0} \phi_{\iota 0}(s) \text{ for } t \in [a_{\iota 0}, b_{\iota 0}] \text{ and a.e. } s \in [0, 1];$$

$$\frac{\partial^i k_\iota}{\partial t^i}(t, s) \geq c_{\iota i} \phi_{\iota i}(s) \text{ for } t \in [a_{\iota i}, b_{\iota i}], \text{ a.e. } s \in [0, 1] \text{ and } i = 1, \ldots, r_\iota.$$

(A3) For $\iota = 1, 2$, $j_\iota = 0, 1, \ldots, r_\iota$, $g_\iota \in L^1([0, 1])$, $g_\iota(t) \geq 0$ a.e. $t \in [0, 1]$, $\phi_{\iota j} \in L^\infty[0, 1]$ and $\int_{a_{\iota j}}^{b_{\iota j}} \phi_{\iota j}(s) g_\iota(s) \, ds > 0$.

Consider the following growth assumptions

(B1) For $\iota = 1, 2$, $l = 0, 1, \ldots, m_\iota$, $j = 0, 1, \ldots, n_\iota$,

$$\limsup_{x_l \to 0, \, y_j \to 0} \max_{t \in [0, 1]} \frac{f_\iota(t, x_0, \ldots, x_{m_\iota}, y_0, \ldots, y_{n_\iota})}{\max\{|x_l|, |y_j|\}} = 0 \qquad (9.5)$$

and

$$\liminf_{x_l \to +\infty, \, y_j \to +\infty} \min_{t \in [0, 1]} \frac{f_\iota(t, x_0, \ldots, x_{m_\iota}, y_0, \ldots, y_{n_\iota})}{\max\{|x_l|, |y_j|\}} = +\infty; \qquad (9.6)$$

(B2) For $\iota = 1, 2$, $l = 0, 1, \ldots, m_\iota$, $j = 0, 1, \ldots, n_\iota$,

$$\liminf_{x_l \to 0,\, y_j \to 0} \min_{t \in [0,1]} \frac{f_\iota(t, x_0, \ldots, x_{m_\iota}, y_0, \ldots, y_{n_\iota})}{\max\{|x_l|, |y_j|\}} = +\infty \qquad (9.7)$$

and

$$\limsup_{x_l \to +\infty,\, y_j \to +\infty} \max_{t \in [0,1]} \frac{f_\iota(t, x_0, x_1, \ldots, x_{m_\iota + n_\iota + 1})}{\max\{|x_l|, |y_j|\}} = 0. \qquad (9.8)$$

9.2 Main result

The main result is given by the next theorem:

Theorem 9.1. *Let, for $\iota = 1, 2$, $f_\iota : [0,1] \times \mathbb{R}^{m_\iota + n_\iota + 2} \to [0, \infty)$ be L^∞-Carathéodory functions such that assumptions (A1)-(A3) hold and or conditions (B1), or (B2), are satisfied. Then problem (9.1) has at least one positive solution $(u, v) \in \big(C^{r_1}[0,1] \times C^{r_2}[0,1] \big)$.*

Proof. Consider the cones K_ι in (9.3), the Banach space $E := K_1 \times K_2$ with the norm given by (9.4), and define the operators $T_1 : E \to K_1$ and $T_2 : E \to K_2$ such that

$$\begin{cases} T_1(u_1, u_2)(t) = \int_0^1 k_1(t, s)\, g_1(s)\, f_1 \begin{pmatrix} s, u_1(s), \ldots, u_1^{(m_1)}(s), \\ u_2(s), \ldots, u_2^{(n_1)}(s) \end{pmatrix} ds \\[3ex] T_2(u_1, u_2)(t) = \int_0^1 k_2(t, s)\, g_2(s)\, f_2 \begin{pmatrix} s, u_1(s), \ldots, u_1^{(m_2)}(s), \\ u_2(s), \ldots, u_2^{(n_2)}(s) \end{pmatrix} ds. \end{cases} \qquad (9.9)$$

The proof will be done in several steps and the main idea is to show that the operator $T : E \to E$ defined by $T = (T_1, T_2)$ has a fixed point on E. For this, according to Lemma 1.5, we need to show that T is completely continuous.

Step 1: $T : E \to E$ *is well defined in E.*

It will be enough to prove that T_ι are well defined in K_ι, for $\iota = 1, 2$. Take $(u_1, u_2) \in E$. By (A2),

$$\|T_1(u_1, u_2)\| = \max_{t \in [0,1]} \int_0^1 k_1(t, s)\, g_1(s)\, f_1 \begin{pmatrix} s, u_1(s), \ldots, u_1^{(m_1)}(s), \\ u_2(s), \ldots, u_2^{(n_1)}(s) \end{pmatrix} ds$$

$$\leq \int_0^1 \phi_{10}(s)\, g_1(s)\, f_1(s, u_1(s), \ldots, u_1^{(m_1)}(s), u_2(s), \ldots, u_2^{(n_1)}(s))\, ds$$

and

$$\min_{t\in[a_{10},b_{10}]} T_1\left(u_1,u_2\right)(t) \geq c_{10}\int_0^1 \phi_{10}(s)g_1(s)\,f_1\left(\begin{array}{c} s,u_1(s),\ldots,u_1^{(m_1)}(s),\\ u_2(s),\ldots,u_2^{(n_1)}(s) \end{array}\right)ds$$

$$\geq c_{10}\|T_1\left(u_1,u_2\right)\|.$$

On the other hand, for $i=1,\ldots,r_1$,

$$\left\|\left(T_1\left(u_1,u_2\right)\right)^{(i)}\right\|$$

$$= \max_{t\in[0,1]}\left|\int_0^1 \frac{\partial^i k_1}{\partial t^i}(t,s)g_1(s)f_1(s,u_1(s),\ldots,u_1^{(m_1)}(s),u_2(s),\ldots,u_2^{(n_1)}(s))\,ds\right|$$

$$\leq \int_0^1 \phi_{1i}(s)g_1(s)f_1(s,u_1(s),\ldots,u_1^{(m_1)}(s),u_2(s),\ldots,u_2^{(n_1)}(s))\,ds,$$

and

$$\min_{t\in[a_{1i},b_{1i}]} \left(T_1\left(u_1,u_2\right)\right)^{(i)}$$

$$= \min_{t\in[a_{1i},b_{1i}]}\int_0^1 \frac{\partial^i k_1}{\partial t^i}(t,s)g_1(s)f_1(s,u_1(s),\ldots,u_1^{(m_1)}(s),u_2(s),\ldots,u_2^{(n_1)}(s))\,ds$$

$$\geq c_{1i}\int_0^1 \phi_{1i}(s)g_1(s)f_1(s,u_1(s),\ldots,u_1^{(m_1)}(s),u_2(s),\ldots,u_2^{(n_1)}(s))\,ds$$

$$\geq c_{1i}\left\|\left(T_1\left(u_1,u_2\right)\right)^{(i)}\right\|.$$

So, for $i=0,1,\ldots,r_1$,

$$\min_{t\in[a_{1i},b_{1i}]} T_1\left(u_1,u_2\right)(t) \geq d_1\|T_1\left(u_1,u_2\right)\|_{C^{r_1}},$$

with $0 < d_1 \leq \max\{c_{1i}, i=0,1,\ldots,r_1\} \leq 1$.

Therefore, $T_1 E \subseteq K_1$. The inclusion $T_2 E \subseteq K_2$ can be proved similarly, and consequently $TE \subset E$.

Step 2: T *is uniformly bounded in* E.

We will prove that T_1 and T_2 are uniformly bounded in K_1 and K_2, respectively.

Consider $(u_1,u_2)\in E$ such that $\|(u_1,u_2)\|_E \leq \rho$, for some $\rho > 0$.

The proof will be done for the operator T_1, as for T_2 the arguments are analogous.

By (A2), (A3) and Definition 9.1,

$$\|T_1\left(u_1,u_2\right)\| = \max_{t\in[0,1]}|T_1\left(u_1,u_2\right)(t)|$$

$$\leq \int_0^1 \phi_{10}(s)g_1(s)f_1(s,u_1(s),\ldots,u_1^{(m_1)}(s),u_2(s),\ldots,u_2^{(n_1)}(s))\,ds$$

$$\leq \int_0^1 \phi_{10}(s)g_1(s)\varphi_\rho(s)ds < +\infty,$$

and, for $i = 1, \ldots, r_1$,

$$\| (T_1 (u_1, u_2))^{(i)} \|$$

$$= \max_{t \in [0,1]} \left| \int_0^1 \frac{\partial^{(i)} k_1}{\partial t^i}(t, s) g_1(s) f_1(s, u_1(s), \ldots, u_1^{(m_1)}(s), u_2(s), \ldots, u_2^{(n_1)}(s)) \, ds \right|$$

$$\leq \int_0^1 \phi_{1i}(s) g_1(s) f_1(s, u_1(s), \ldots, u_1^{(m_1)}(s), u_2(s), \ldots, u_2^{(n_1)}(s)) \, ds$$

$$\leq \int_0^1 \phi_{1i}(s) g_1(s) \varphi_\rho(s) ds < +\infty.$$

Therefore $\|T_1 (u_1, u_2)\|_{C^{r_1}} < +\infty$, and, so, T_1 is uniformly bounded in K_1.

By an analogous method it can be proved that T_2 is uniformly bounded in K_2, and, therefore, T is uniformly bounded in E.

Step 3: *T is equicontinuous in E.*

This step will be shown if T_1 and T_2 are equicontinuous in K_1 and K_2, respectively. The calculus will be done only for T_1, as the other case is similar.

Consider $t_1, t_2 \in [0,1]$. By (A1),

$$|T_1 (u_1, u_2) (t_1) - T_1 (u_1, u_2) (t_2)|$$

$$\leq \int_0^1 |k_1(t_1, s) - k_1(t_2, s)| \, g_1(s) \, f_1 \begin{pmatrix} s, u_1(s), \ldots, u_1^{(m_1)}(s), \\ u_2(s), \ldots, u_2^{(n_1)}(s) \end{pmatrix} ds$$

$$\leq \int_0^1 |k_1(t_1, s) - k_1(t_2, s)| \, g_1(s) \varphi_\rho(s) ds \to 0 \quad \text{as} \quad t_1 \to t_2,$$

and, for $i = 1, \ldots, r_1$,

$$|(T_1 (u_1, u_2))^{(i)}(t_1) - (T_1 (u_1, u_2))^{(i)}(t_2)|$$

$$\leq \int_0^1 \left| \frac{\partial^i k_1}{\partial t^i}(t_1, s) - \frac{\partial^i k_1}{\partial t^i}(t_2, s) \right| g_1(s) \, f_1 \begin{pmatrix} s, u_1(s), \ldots, u_1^{(m_1)}(s), \\ u_2(s), \ldots, u_2^{(n_1)}(s) \end{pmatrix} ds$$

$$\leq \int_0^1 \left| \frac{\partial^i k_1}{\partial t^i}(t_1, s) - \frac{\partial^i k_1}{\partial t^i}(t_2, s) \right| g_1(s) \varphi_\rho(s) ds \to 0 \quad \text{as} \quad t_1 \to t_2.$$

Therefore, T_1 is equicontinuous in K_1.

By the same way it can be proved that T_2 is equicontinuous in K_2. Then T is equicontinuous in E.

By the Arzelà-Ascoli Theorem, T is completely continuous in E. Assume that condition (B1) holds.

Step 4: $\|T(u_1, u_2)\|_E \leq \|(u_1, u_2)\|_E$, for $(u_1, u_2) \in E \cap \partial\Omega_1$ with $\Omega_1 = \{(u_1, u_2) \in E : \|(u_1, u_2)\|_E < \rho_1\}$, for some $\rho_1 > 0$.

To prove that

$$\max\{\|T_1(u_1, u_2)\|_{C^{r_1}}, \|T_2(u_1, u_2)\|_{C^{r_2}}\} \leq \|(u_1, u_2)\|_E,$$

it will be enough to show that

$$\|T_1(u_1, u_2)\|_{C^{r_1}} \leq \|(u_1, u_2)\|_E \text{ and } \|T_2(u_1, u_2)\|_{C^{r_2}} \leq \|(u_1, u_2)\|_E.$$

As $(u_1, u_2) \in E \cap \partial\Omega_1$ then $\|(u_1, u_2)\|_E = \rho_1$.
For $i = 0, 1, \ldots, r_1$, and (A3), let us define

$$\varepsilon_1 := \min\left\{\frac{1}{\int_0^1 \phi_{1i}(s)g_1(s)ds}\right\}. \tag{9.10}$$

By (9.5), there exists $0 < \rho_1 < 1$ such that

$$f_1(t, u_1(t), \ldots, u_1^{(m_1)}(t), u_2(t), \ldots, u_2^{(n_1)}(t)) \leq \varepsilon_1 \|(u_1, u_2)\|_E, \tag{9.11}$$

for $\|(u_1, u_2)\|_E \leq \rho_1$.
By (A2), (10.11), and (10.10),

$$T_1(u_1, u_2)(t) = \int_0^1 k_1(t, s)g_1(s)f_1(s, u_1(s), \ldots, u_1^{(m_1)}(s), u_2(s), \ldots, u_2^{(n_1)}(s)) ds$$

$$\leq \int_0^1 \phi_{10}(s)g_1(s)\varepsilon_1 \|(u_1, u_2)\|_E \, ds$$

$$= \varepsilon_1 \rho_1 \int_0^1 \phi_{10}(s)g_1(s)ds < \rho_1 = \|(u_1, u_2)\|_E,$$

and, for $i = 1, \ldots, r_1$,

$$(T_1(u_1, u_2)(t))^{(i)} = \int_0^1 \frac{\partial^i k_1}{\partial t^i}(t, s)g_1(s) f_i\left(\begin{array}{c} s, u_1(s), \ldots, u_1^{(m_1)}(s), \\ u_2(s), \ldots, u_2^{(n_1)}(s) \end{array}\right) ds$$

$$\leq \int_0^1 \phi_{1i}(s)g_1(s)\varepsilon_1 \|(u_1, u_2)\|_E \, ds < \rho_1 = \|(u_1, u_2)\|_E.$$

Ergo, $\|T_1(u_1, u_2)\|_{C^{r_1}} \leq \|(u_1, u_2)\|_E$, for $(u_1, u_2) \in E \cap \partial\Omega_1$. By similar calculations it can be proved that $\|T_2(u_1, u_2)\|_{C^{r_2}} \leq \|(u_1, u_2)\|_E$ and therefore, $\|T(u_1, u_2)\|_E \leq \|(u_1, u_2)\|_E$, for $(u_1, u_2) \in E \cap \partial\Omega_1$.

Step 5: $\|T(u_1, u_2)\|_E \geq \|(u_1, u_2)\|_E$, for $u \in E \cap \partial\Omega_2$ with $\Omega_2 = \{(u_1, u_2) \in E : \|(u_1, u_2)\|_E < \rho_2\}$, for some $\rho_2 > 0$.

If there is $i_0 \in \{0, 1, \ldots, m_1\}$, or $j_0 \in \{0, 1, \ldots, n_1\}$, such that $u_1^{(i_0)}(t) \to +\infty$ and $u_2^{(j_0)}(t) \to +\infty$, then $\|(u_1, u_2)\|_E \to +\infty$.

By (9.6), for $\iota = 1, 2$, there exist $\rho_\iota^* > 0$ and $\theta > 0$, such that, when $\| (u_1, u_2) \|_E \geq \theta$ we have

$$f_1(t, u_1(t), \ldots, u_1^{(m_1)}(t), u_2(t), \ldots, u_2^{(n_1)}(t)) \geq \| (u_1, u_2) \|_E. \qquad (9.12)$$

Define, for $i = 0, 1, \ldots, r_1$,

$$\xi_1 := \max \left\{ \frac{1}{c_{1i} \int_0^1 \phi_{1i}(s) g_1(s) ds} \right\}. \qquad (9.13)$$

Let $(u_1, u_2) \in E$ be such that $\| (u_1, u_2) \|_E = \rho_2$, with $\rho_2 > \rho_1$. Now from (A2), and (9.13),

$$\begin{aligned}
T_1(u_1, u_2)(t) &\geq \int_{a_{10}}^{b_{10}} k_1(t, s) g_1(s) f_1 \begin{pmatrix} s, u_1(s), \ldots, u_1^{(m_1)}(s), \\ u_2(s), \ldots, u_2^{(n_1)}(s) \end{pmatrix} ds \\
&\geq c_{10} \int_{a_{10}}^{b_{10}} \phi_{10}(s) g_1(s) f_1 \begin{pmatrix} s, u_1(s), \ldots, u_1^{(m_1)}(s), \\ u_2(s), \ldots, u_2^{(n_1)}(s) \end{pmatrix} ds \\
&\geq c_{10} \int_{a_{10}}^{b_{10}} \phi_{10}(s) g_1(s) \xi_1 \| (u_1, u_2) \|_E \, ds \\
&= c_{10} \xi_1 \rho_2 \int_{a_{10}}^{b_{10}} \phi_{10}(s) g_1(s) ds \geq \rho_2 = \| (u_1, u_2) \|_E,
\end{aligned}$$

and analogously, for $i = 1, \ldots, r_1$,

$$\begin{aligned}
(T_1(u_1, u_2)(t))^{(i)} &\geq \int_{a_{1i}}^{b_{1i}} \frac{\partial^{(i)} k_1}{\partial t^i}(t, s) g_1(s) f_1 \begin{pmatrix} s, u_1(s), \ldots, u_1^{(m_1)}(s), \\ u_2(s), \ldots, u_2^{(n_1)}(s) \end{pmatrix} ds \\
&\geq c_{1i} \int_{a_{1i}}^{b_{1i}} \phi_{1i}(s) g_1(s) f_1 \begin{pmatrix} s, u_1(s), \ldots, u_1^{(m_1)}(s), \\ u_2(s), \ldots, u_2^{(n_1)}(s) \end{pmatrix} ds \\
&\geq c_{1i} \xi_1 \rho_2 \int_{a_{1i}}^{b_{1i}} \phi_{1i}(s) g_1(s) \, ds \geq \rho_2 = \| (u_1, u_2) \|_E.
\end{aligned}$$

Therefore, $\| T_1(u_1, u_2) \|_{C^{r_1}} \geq \| (u_1, u_2) \|_E$, for $(u_1, u_2) \in E \cap \partial \Omega_2$. Analogously it can be shown that $\| T_2(u_1, u_2) \|_{C^{r_2}} \geq \| (u_1, u_2) \|_E$, for $(u_1, u_2) \in E \cap \partial \Omega_2$, and, therefore, $\| T(u_1, u_2) \|_E \geq \| (u_1, u_2) \|_E$, for $(u_1, u_2) \in E \cap \partial \Omega_2$.

By Lemma 1.5, the operator T has a fixed point in $K \cap (\overline{\Omega_2} \backslash \Omega_1)$ which in turn is a solution of our problem.

Now assume that (B2) holds.

Step 6: $\| T(u_1, u_2) \|_E \geq \| (u_1, u_2) \|_E$, for $u \in E \cap \partial \Omega_3$ with $\Omega_3 = \{ (u_1, u_2) \in E : \| (u_1, u_2) \|_E < \rho_3 \}$, for some $\rho_3 > 0$.

Taking $\xi_1 > 0$ as in (9.13), we see that by (9.7) there exists $0 < \rho_3 < 1$ such that

$$(t, u_1(t), \ldots, u_1^{(m_1)}(t), u_2(t), \ldots, u_2^{(n_1)}(t)) \in [0, 1] \times [0, \rho_{3_t}]^{m_1+n_1+2}$$

and

$$f_1(t, u_1(t), \ldots, u_1^{(m_1)}(t), u_2(t), \ldots, u_2^{(n_1)}(t)) \geq \xi_t \| (u_1, u_2) \|_E. \qquad (9.14)$$

Consider $(u_1, u_2) \in E$ such that $\| (u_1, u_2) \|_E = \rho_{3_t}$. Then, applying similar inequalities as in Step 5, we obtain that $\|T(u_1, u_2)\|_E \geq \| (u_1, u_2) \|_E$.

Step 7: $\|T(u_1, u_2)\|_E \leq \| (u_1, u_2) \|_E$, for $(u_1, u_2) \in E \cap \partial\Omega_4$ with $\Omega_4 = \{(u_1, u_2) \in E : \| (u_1, u_2) \|_E < \rho_4\}$, for some $\rho_4 > 0$.

Case 7.1. *Suppose that f_1 is bounded.*

Then there is an $N > 0$ such that $f_1(t, u_1(t), \ldots, u_1^{(m_1)}(t), u_2(t), \ldots,$ $u_2^{(n_1)}(t)) \leq N$ for all $(t, u_1(t), \ldots, u_1^{(m_1)}(t), u_2(t), \ldots, u_2^{(n_1)}(t)) \in [0, 1] \times [0, +\infty)^{m_1+n_1+2}$. Choose

$$\rho_4 := \max \left\{ \rho_3 + 1, N \int_0^1 \phi_{1i}(s)g_1(s)ds : i = 0, 1, \ldots, r_1 \right\}$$

and take $(u_1, u_2) \in E$ with $\| (u_1, u_2) \|_E = \rho_4$. Then,

$$T_1(u_1, u_2)(t) = \int_0^1 k_1(t, s)g_1(s)f_1(s, u_1(s), \ldots, u_1^{(m_1)}(s), u_2(s), \ldots, u_2^{(n_1)}(s))ds$$

$$\leq N \int_0^1 \phi_{10}(s)g_1(s)ds \leq \rho_4, \text{ for } t \in [0, 1],$$

and for $i = 1, \ldots, r_1$,

$$(T_1(u_1, u_2)(t))^{(i)} = \int_0^1 \frac{\partial^i k_1}{\partial t^i}(t, s)g_1(s)f\, f_1 \left(\begin{array}{c} s, u_1(s), \ldots, u_1^{(m_1)}(s), \\ u_2(s), \ldots, u_2^{(n_1)}(s) \end{array} \right) ds$$

$$\leq N \int_0^1 \phi_{1i}(s)g_1(s)ds \leq \rho_4, \text{ for } t \in [0, 1].$$

Thus, $\|T_1(u_1, u_2)\|_{C^{r_1}} \leq \| (u_1, u_2) \|_E$.

The same arguments can be applied to show that $\|T_2(u_1, u_2)\|_{C^{r_2}} \leq \| (u_1, u_2) \|_E$. So, $\|T(u_1, u_2)\|_E \leq \| (u_1, u_2) \|_E$.

Case 7.2. *Suppose that f_1 is unbounded.*

By (9.8), there exists $\mu > 0$ such that

$$\max \left\{ \mu \int_0^1 \phi_{1i}(s)g_1(s)ds : i = 0, 1, \ldots, r_1 \right\} \leq 1 \qquad (9.15)$$

and

$$f_1(t, u_1(t), \ldots, u_1^{(m_1)}(t), u_2(t), \ldots, u_2^{(n_1)}(t)) \leq \mu \| (u_1, u_2) \|_E, \qquad (9.16)$$

for every $M > 0$ such that $\| (u_1, u_2) \|_E \geq M$.

Define

$$\rho_4 := \max\{M, \rho_3 + 1\}.$$

Then, for $(u_1, u_2) \in E \cap \partial\Omega_4$ we have $\| (u_1, u_2) \|_E = \rho_4$ and, by (9.16),

$$f_1(t, u_1(t), \ldots, u_1^{(m_1)}(t), u_2(t), \ldots, u_2^{(n_1)}(t)) \leq \mu \| (u_1, u_2) \|_E \leq \mu\rho_4. \quad (9.17)$$

So,

$$T_1(u_1, u_2)(t) \leq \int_0^1 \phi_{10}(s)g_1(s)\mu\rho_4 \, ds \leq \mu\rho_4 \int_0^1 \phi_{10}(s)g_1(s) \, ds \leq \rho_4,$$

and for $i = 1, \ldots, r_1$,

$$(T_1(u_1, u_2)(t))^{(i)} \leq \int_0^1 \phi_{1i}(s)g_1(s)\mu\rho_4 \, ds \leq \mu\rho_4 \int_0^1 \phi_{1i}(s)g_1(s) \, ds \leq \rho_4.$$

Therefore, $\|T_1(u_1, u_2)\|_{C^{r_1}} \leq \| (u_1, u_2) \|_E$, for $(u_1, u_2) \in E \cap \partial\Omega_4$.

In the same way we can have $\|T_2(u_1, u_2)\|_{C^{r_2}} \leq \| (u_1, u_2) \|_E$, for $(u_1, u_2) \in E \cap \partial\Omega_4$, and, therefore, $\|T(u_1, u_2)\|_E \leq \| (u_1, u_2) \|_E$, for $(u_1, u_2) \in E \cap \partial\Omega_4$.

The remaining cases (f_2 bounded or unbounded) can be processed by similar techniques.

By Lemma 1.5, the operator T has a fixed point in $E \cap (\overline{\Omega_4} \backslash \Omega_3)$ that, in turn, is a solution of the problem. $\qquad \square$

9.3   Example

Consider the following coupled system composed by third and second order nonlinear equations, with three-point boundary conditions

$$\begin{cases} -u_1'''(t) = (t^2 + 1)\left(e^{-(u_2'(t)+u_1(t))^2} + \sqrt{|u_1'(t) + u_2(t)|}\right) \\[2mm] u_2''(t) = t^4\,(2 + \cos(u_2(t) + u_1(t)))^2\,(\sin(u_1'(t)u_2'(t)) + 1) \\[2mm] u(0) = u'(0) = 0,\ u'(1) = \frac{3}{2}u'\left(\frac{1}{2}\right) \\[2mm] v(0) = 0,\ v'(1) = \frac{3}{2}v'\left(\frac{1}{2}\right). \end{cases} \qquad (9.18)$$

Note that the problem (9.18) can be rewritten as the following system of integral equations

$$\begin{cases} u_1(t) = \int_0^1 k_1(t,s) \ (s^2+1) \left(e^{-(u_2'(s)+u_1(s))^2} + \sqrt{|u_1'(s) + u_2(s)|} \right) ds, \\[2mm] u_2(t) = \int_0^1 k_2(t,s) \ s^4 \ (2 + \cos(u_2(s) + u_1(s)))^2 \ (\sin(u_1'(s)u_2'(s)) + 1) \, ds, \end{cases}$$
(9.19)

where the kernel functions $k_1(t,s)$ and $k_2(t,s)$ are given by the corresponding Green's functions

$$G_1(t,s) = \begin{cases} ts - \frac{s^2}{2} + 2t^2 s, & s \le \min\{\frac{1}{2}, t\}, \\[1mm] \frac{t^2}{2} + 2t^2 s, & t \le s \le \frac{1}{2}, \\[1mm] ts - \frac{s^2}{2} + t^2 \left(\frac{3}{2} - 2s \right), & \frac{1}{2} \le s \le t, \\[1mm] 2t^2(1-s), & \max\{\frac{1}{2}, t\} \le s, \end{cases}$$

$$G_2(t,s) = \begin{cases} t - s, & s \le t \le 1, \\[1mm] 2t, & \frac{1}{2} \le t \le s \le 1, \end{cases}$$

respectively.

Clearly system (9.19) is a particular case of (9.1) with $r_1 = r_2 = m_1 = m_2 = n_1 = n_2 = 1$, $g_1(t) = t^2 + 1$, $g_2(t) = t^4$, $k_1(t,s) = G_1(t,s)$, $k_2(t,s) = G_2(t,s)$, and

$$f_1(t, x, y, z, w) = \left(e^{-(w+x)^2} + \sqrt{|y+z|} \right),$$

$$f_2(t, x, y, z, w) = (2 + \cos(z+x))^2 \ (\sin(yw) + 1) \, .$$

These functions $f_1, f_2 : \mathbb{R}^5 \to [0, \infty)$ are L^∞−Carathéodory as, for $\rho > 0$, when $\max\{|x|, |y|, |z|, |w|\} < \rho$, there exist functions $\varphi_{1\rho}, \varphi_{2\rho} \in L^\infty([0,1])$ such that

$$f_1(t, x, y, z, w) \le \left(1 + \sqrt{|2\rho|} \right) := \varphi_{1\rho}$$

$$f_2(t, x, y, z, w) = (2 + \cos(z+x))^2 \ (\sin(yw) + 1) \le 18 := \varphi_2.$$

The first derivative of the Green's functions are positive, with

$$\frac{\partial G_1}{\partial t}(t,s) = \begin{cases} s + 4ts, & s \le \min\{\frac{1}{2}, t\}, \\[1mm] t + 4ts, & t \le s \le \frac{1}{2}, \\[1mm] t + 2t \left(\frac{3}{2} - 2s \right), & \frac{1}{2} \le s \le t, \\[1mm] 4t(1-s), & \max\{\frac{1}{2}, t\} \le s, \end{cases}$$

$$\frac{\partial G_2}{\partial t}(t,s) = \begin{cases} 1, & s \le t \le 1, \\ 2, & \frac{1}{2} \le t \le s \le 1. \end{cases}$$

Therefore (A1) holds and, to show that (A2) is verified, we follow the arguments in [100] (Lemma 4.1 – Lemma 4.4), for G_1 and any $(t,s) \in [0,1] \times [0,1]$, to obtain

$$0 \le G_1(t,s) \le 10s(1-s) := \phi_{10}(s), \quad 0 \le \frac{\partial G_1}{\partial t}(t,s) \le 4(1-s) := \phi_{11}(s)$$

and for $(t,s) \in [\frac{1}{3}, \frac{1}{2}] \times [0,1]$, it follows that

$$c_{10} = \frac{1}{90}, \quad c_{11} = \frac{3}{4}$$

and

$$G_1(t,s) \ge c_{10}\phi_{10}(s) = \frac{1}{9}s(1-s), \quad \frac{\partial G_1}{\partial t}(t,s) \ge c_{11}\phi_{11}(s) = 3(1-s),$$

and for G_2, taking

$$\phi_{20}(s) = 2, \quad c_{20} = \frac{1}{20}, \quad \phi_{21}(s) = \frac{5}{2}, \quad c_{21} = \frac{1}{5},$$

pursue that

$$G_2(t,s) \ge c_{20}\phi_{20}(s) = \frac{1}{10}, \quad \frac{\partial G_2}{\partial t}(t,s) \ge c_{21}\phi_{21}(s) = \frac{1}{2}.$$

Condition (A3) is satisfied as the referred four integrals are trivially positive.

For $j = 0, 1$, we have

$$\limsup_{u_1^{(j)}, u_2^{(j)} \to 0} \max_{t \in [0,1]} \frac{\left(e^{-(u_2'(t)+u_1(t))^2} + \sqrt{|u_1'(t) + u_2(t)|}\right)}{\max\left\{\left|u_1^{(j)}\right|, \left|u_2^{(j)}\right|\right\}} = +\infty,$$

$$\liminf_{u_1^{(j)}, u_2^{(j)} \to +\infty} \min_{t \in [0,1]} \frac{\left(e^{-(u_2'(t)+u_1(t))^2} + \sqrt{|u_1'(t) + u_2(t)|}\right)}{\max\left\{\left|u_1^{(j)}\right|, \left|u_2^{(j)}\right|\right\}} = 0,$$

$$\liminf_{u_1^{(j)}, u_2^{(j)} \to 0} \min_{t \in [0,1]} \frac{(2 + \cos(u_2(t) + u_1(t)))^2 (\sin(u_1'(t)u_2'(t)) + 1)}{\max\left\{\left|u_1^{(j)}\right|, \left|u_2^{(j)}\right|\right\}} = +\infty,$$

$$\limsup_{u_1^{(j)}, u_2^{(j)} \to +\infty} \max_{t \in [0,1]} \frac{(2 + \cos(u_2(t) + u_1(t)))^2 (\sin(u_1'(t)u_2'(t)) + 1)}{\max\left\{\left|u_1^{(j)}\right|, \left|u_2^{(j)}\right|\right\}} = 0,$$

and, therefore, conditions (B1) hold.

So, by Theorem 11.1, there is at least one positive solution $(u,v) \in \left(C^1[0,1] \times C^1[0,1]\right)$ of problem (9.19), which is the solution of (9.18).

Chapter 10

Coupled Hammerstein systems with sign-changing kernels

Hammerstein integral equations have been extensively studied under several topics such as: the existence, non existence and multiplicity of solutions, and applications to real phenomena, among others, as it can be seen in $[3, 4, 13, 33, 37, 45, 68, 69, 200, 235, 268]$, and the references therein.

Like many relevant theories in mathematics, especially in topology and functional analysis, the theory of integral equations has been and continues to be a field of many research and applications. These equations are especially relevant in physics and are often used to reformulate or rewrite mathematical problems.

Hammerstein equations are special subclasses of nonlinear integral equations of Fredholm type

$$u(x) = g(x) + \int_a^b k(x, y, u(x), u(y)) f(y, u(y)) dy, \quad x \in [a, b],$$

and their study was initiated by Hammerstein (see $[27, 110]$). They appear in some mathematical models such as: electrostatic drift waves, low-frequency electromagnetic perturbation (see $[88]$), signal theory (see $[154]$), nanocantilever beams (see $[223, 224]$) and conservation laws (see $[70, 71]$).

The existence, uniqueness, multiplicity, positivity and localization of solutions are the most studied and predominant elements about Hammerstein integral equations. Citing just a few examples in the literature, in $[265]$, the authors use fixed point index theory to establish their main result, based on *a priori* estimates, achieved by nonnegative matrices; in $[67]$, Coclite studies the existence of a positive measurable solution of the Hammerstein equation of the first kind, with a singular nonlinear term at the origin; in $[53]$, to prove the existence of positive solutions for systems of nonlinear Hammerstein integral equations, the authors present new criteria on the existence of fixed points that combine some monotonicity assumptions with the classical fixed point index theory; in [58], Cardinali *et al.*, examine multivalued Hammerstein integral equations defined in a separable reflexive Banach space. They prove existence theorems for convex and nonconvex problems; in [87], the researchers study solutions of the nonlinear Hammerstein integral equation with sign-changing kernels by using Ricceri's variational principle and critical point theory techniques. They combine the effects of sublinear and superlinear terms to establish new existence and multiplicity results; in [272], the researchers study the existence and the uniqueness of iterative positive solutions for a class of nonlinear singular integral equations in which the nonlinear terms may be singular in both variables: time and space. By using the fixed point theorem of mixed monotone operators in cones, they establish the conditions for the existence and uniqueness of positive solutions to the problem; in [66], the authors rewrite a fourth-order boundary value problem as a perturbed Hammerstein integral equation of the form

$$u(t) = \gamma(t)h(u(1)) + \int_0^1 k(t,s)f(s,u(s))ds,$$

where the perturbation is related with nonlocal boundary conditions, the kernel function $k : [0,1] \times [0,1] \to (0,+\infty)$ is measurable, positive and verifies adequate regularity conditions. Applying topological methods, some results on the existence, non-existence, localization and multiplicity of nontrivial solutions are obtained; in [100], Graef *et al.*, study the Hammerstein generalized integral equation

$$u(t) = \int_0^1 k(t,s)\ g(s)\ f(s,u(s),u'(s),\dots,u^{(m)}(s))\,ds, \qquad (10.1)$$

where $k : [0,1]^2 \to \mathbb{R}$ is a kernel function such that $k \in W^{m,1}\left([0,1]^2\right)$, m is a positive integer with $m \geq 1$, $g : [0,1] \to [0,\infty)$ with $g(t) \geq 0$ a.e. $t \in [0,1]$, and $f : [0,1] \times \mathbb{R}^{m+1} \to [0,\infty)$ is a $L^\infty-$Carathéodory function. The existence of solutions of (10.1) will be obtained here via the well-known Krasnosel'skiĭ-Guo cone compression/expansion fixed point theorem. It is pointed out that the kernels, and their partial derivatives with respect to the first variable, may be discontinuous and may change sign since they are only required to be positive on some subsets of $[0,1]$ of positive measure.

More specifically, about Hammerstein-type coupled systems of integral equations, in [72], by constructing a special cone and using fixed point index theory, Cui and Sun, investigate the existence of positive solutions of singular superlinear coupled integral boundary value problems for differential systems

$$\begin{cases} -x''(t) = f_1(t, x(t), y(t)) \\ -y''(t) = f_2(t, x(t), y(t)), t \in (0, 1), \\ \quad x(0) = y(0) = 0, \\ \quad x(1) = \alpha[y], \\ \quad y(1) = \beta[x], \end{cases}$$

where $f_1, f_2 : (0, 1) \times [0, +\infty)^2 \to [0, +\infty)$ are continuous and may be singular at $t = 0, 1$, and $\alpha[x], \beta[x]$ are bounded linear functionals on $C[0, 1]$.

In [259], the authors study the existence and multiplicity of positive solutions for the system of nonlinear Hammerstein integral equations

$$u(x) = \int_0^1 k_1(x, y) \, f_1(y, u(y), v(y), w(y)) \, dy,$$

$$v(x) = \int_0^1 k_2(x, y) \, f_2(y, u(y), v(y), w(y)) \, dy,$$

$$w(x) = \int_0^1 k_2(x, y) \, f_3(y, u(y), v(y), w(y)) \, dy,$$

where $k \in C\left([0, 1] \times [0, 1], \mathbb{R}_+\right)$, f_1, f_2, $f_3 \in C\left([0, 1] \times \mathbb{R}_+^3, \mathbb{R}_+\right)$. The authors use the concave functions to characterize growing and interacting behaviors of nonlinearities f_1, f_2, f_3, so that they cover three cases: first, with all superlinear, second, with all sublinear and the last with two superlinear and the other sublinear. The arguments are based on *a priori* estimates achieved by Jensen's integral inequality for concave functions.

Recently, in [131], Infante and Minhós, extending the results in [200], to prove the existence, multiplicity and non-existence results for nontrivial solutions of the systems

$$\begin{cases} u(t) = \int_0^1 k_1(t, s) \, g_1(s) \, f_1(s, u(s), u'(s), v(s), v'(s)) \, ds, \\ v(t) = \int_0^1 k_2(t, s) \, g_2(s) \, f_2(s, u(s), u'(s), v(s), v'(s)) \, ds, \end{cases}$$

obtained via fixed index theorem and cones theory.

Motivated by these works we consider the following generalized coupled systems of integral equations of Hammerstein-type

$$\begin{cases} u_1(t) = \int_0^1 k_1(t, s) \, g_1(s) \, f_1(s, u_1(s), \dots, u_1^{(m_1)}(s), u_2(s), \dots, u_2^{(n_1)}(s)) \, ds, \\ u_2(t) = \int_0^1 k_2(t, s) \, g_2(s) \, f_2(s, u_1(s), \dots, u_1^{(m_2)}(s), u_2(s), \dots, u_2^{(n_2)}(s)) \, ds, \end{cases}$$

$$\tag{10.2}$$

where $k_\iota : [0, 1]^2 \to \mathbb{R}$, $\iota = 1, 2$, are the kernel functions such that $k_\iota \in W^{r_\iota, 1}\left([0, 1]^2\right)$, $r_1 = \max\{m_1, m_2\}$, $r_2 = \max\{n_1, n_2\}$, with $m_\iota, n_\iota \geq 0$,

$g_\iota \in L^1([0,1])$ with $g_\iota(t) \geq 0$ for a.e. $t \in [0,1]$, and $f_\iota : [0,1] \times \mathbb{R}^{m_\iota + n_\iota + 2} \to [0, \infty)$ are L^∞−Carathéodory functions.

The main purpose of this chapter is to overcome the positivity of Green's functions and their derivatives for high-order coupled systems and consequently to extend the number of real applications of these results. In this way, this work provides new features:

- The coupled system of Hammerstein-type integral equations contains nonlinearities depending on several derivatives of both variables and, moreover, the derivatives can be of different order on each variable and each equation, which increases the range of applications.
- A new type of cone is introduced, where some requirements may be satisfied only on some subintervals of the domain.
- The kernel functions can change sign, ensuring the positivity only on some intervals, eventually degenerate, that is, having only one point. This is overcome with a convenient fixed point theorem on cone theory (see Theorem 10.1).

For the reader's convenience let us remark that how this chapter improves and generalizes [131], even for the cases $m_1 = n_1 = m_2 = n_2 = 1$, and the integral system is related to a second-order system of ODE:

Assume also that the kernel functions $k_\iota(t,s)$, $\iota = 1, 2$, are given by the corresponding Green's functions $G_1(x,s)$ and $G_2(x,s)$. In this case $\frac{\partial G_\iota}{\partial x}(x,s)$, $\iota = 1, 2$, do not verify the assumption (A2) of [131] as

$$\lim_{t \to \tau} \left| \frac{\partial^i k_\iota}{\partial t^i}(t,s) - \frac{\partial^i k_\iota}{\partial t^i}(\tau, s) \right| \neq 0$$

for τ in the diagonal of the square, that is for (τ, τ).

However $\frac{\partial G_\iota}{\partial x}(x,s)$, $\iota = 1, 2$, verify the assumption (A1) of our work as the jump is equal to 1. So for second-order, our result is more general than [131] and allows to consider the Green's functions to play the role of the kernel functions $k_\iota(t,s)$, $\iota = 1, 2$. For third-order, assumption (A2) of [131] and our assumption (A1) are not equivalent but they cover the same cases for $m_1 = n_1 = m_2 = n_2 = 1$. Another important issue is the following: In our work the positivity of kernel functions $k_\iota(t,s)$ or their derivatives, given by our (A2), can be verified in sets of null measure, that is $[q_{\iota i}, p_{\iota i}]$, with $q_{\iota i} = p_{\iota i}$. However the inequality verified by the kernel function, given by (A3) of [131], must hold on sets of positive measure.

10.1 Hypothesis and backgrounds

The aim of our assumptions is to ensure the nonnegativity of the kernel functions that occur in some subintervals of $J := [0, 1]$, or even at a single point.

The regularity of the nonlinear functions f_1 and f_2 is given by the L^∞–Carathéodory functions from Definition 9.1.

The Guo-Krasnoselskii fixed point existence theorem, for expansive and compressive cones theory, will be a key tool in the arguments.

Theorem 10.1 ([107, Theorem 2.3.3]). *Let P be a cone of real Banach space E and Ω_1, Ω_2 two open sets of E such that $0 \in \Omega_1$, $\overline{\Omega_1} \subset \Omega_2$. If $T : P \cap (\overline{\Omega_2} \setminus \Omega_1) \to P$ is a completely continuous operator such that either*

(i) $Tu \not\geq u$, $\forall u \in P \cap \partial\Omega_1$, *and* $Tu \not\leq u$, $\forall u \in P \cap \partial\Omega_2$, *or*
(ii) $Tu \not\leq u$, $\forall u \in P \cap \partial\Omega_1$, *and* $Tu \not\geq u$, $\forall u \in P \cap \partial\Omega_2$,

is satisfied. Then T has a fixed point in $P \cap (\Omega_2 \setminus \overline{\Omega_1})$.

In this chapter we consider the following assumptions:

(A1) For $\iota = 1, 2$, the functions are such that $k_\iota : J^2 \to \mathbb{R}$, $k_\iota \in W^{r_\iota, 1}(J^2)$. Moreover, for $i = 0, \ldots, r_\iota - 1$, with $r_\iota > 1$, it holds that for every $\varepsilon > 0$ and every fixed $\tau \in [0, 1]$, there exists some $\delta(\tau) > 0$ such that $|t - \tau| < \delta(\tau)$ implies that

$$\left| \frac{\partial^i k_\iota}{\partial t^i}(t, s) - \frac{\partial^i k_\iota}{\partial t^i}(\tau, s) \right| < \varepsilon, \quad \text{for a.e. } s \in [0, 1],$$

and for the r_ι-th derivative of k_ι, $|t - \tau| < \delta(\tau)$ implies that

$$\left| \frac{\partial^{r_\iota} k_\iota}{\partial t^{r_\iota}}(t, s) - \frac{\partial^{r_\iota} k_\iota}{\partial t^{r_\iota}}(\tau, s) \right| < \varepsilon,$$

for a.e. $s < \min\{t, \tau\}$ and for a.e. $s > \max\{t, \tau\}$.

(A2) For $\iota = 1, 2$, there is $I_{\iota 0} \neq \varnothing$ such that, for each $i \in I_{\iota 0} \subseteq I_\iota := \{0, 1, \ldots, r_\iota\}$, there exist subintervals $[q_{\iota i}, p_{\iota i}]$ such that

$$\frac{\partial^i k_\iota}{\partial t^i}(t, s) \geq 0, \quad \text{for all } t \in [q_{\iota i}, p_{\iota i}], \ s \in I_{\iota 0}.$$

Eventually, these intervals could be degenerated, that is $q_{\iota i} = p_{\iota i}$.

(A3) For $\iota = 1, 2$, there are positive functions $\phi_{\iota j} \in L^\infty(J)$, $j \in I_\iota$, such that

$$\left| \frac{\partial^j k_\iota}{\partial t^j}(t, s) \right| \leq \phi_{\iota j} \text{ for } t \in J \text{ and a.e. } s \in J.$$

(A4) For $\iota = 1, 2$, there is $I_{\iota 1} \neq \varnothing$ such that, for each $j \in I_{\iota 1} \subseteq I_{\iota 0}$, there exist subintervals $[a_{\iota j}, b_{\iota j}]$, $[c_{\iota j}, d_{\iota j}] \subseteq J$, non negative functions $\psi_{\iota j} : J \to [0, \infty)$, and constants $\xi_{\iota j} \in (0, 1)$, such that:

$$\left| \frac{\partial^j k_\iota}{\partial t^j}(t, s) \right| \leq \psi_{\iota j}(s) \text{ for } t \in [c_{\iota j}, d_{\iota j}] \text{ and a.e. } s \in J;$$

$$\frac{\partial^j k_\iota}{\partial t^j}(t, s) \geq \xi_{\iota j} \psi_{\iota j}(s), \text{ for } t \in [a_{\iota j}, b_{\iota j}] \text{ and a.e. } s \in J.$$

Moreover, $\psi_{\iota j} g_\iota \in L^1(J)$ verifies

$$\int_{a_{\iota j}}^{b_{\iota j}} \psi_{\iota j}(s)\, g_\iota(s)\, ds > 0.$$

(A5) For $\iota = 1, 2$, there are $i_0 \in I_{\iota 0}$ such that $\{0, 1, ..., i_0\} \subseteq I_{\iota 0}$ and either $[c_{\iota i_0}, d_{\iota i_0}] \equiv J$ or $[q_{\iota i_0}, p_{\iota i_0}] \equiv J$.

Taking into account the above properties we introduce, for $\iota = 1, 2$, the following cones

$$K_\iota := \left\{ \begin{array}{l} w \in C^{r_\iota}(J, \mathbb{R}) : w^{(i)}(t) \geq 0,\, t \in [q_{\iota i}, p_{\iota i}],\, i \in I_{\iota 0}, \\[2mm] \min\limits_{t \in [a_{\iota j}, b_{\iota j}]} w^{(j)}(t) \geq \xi_{\iota j} \|w^{(j)}\|_{[c_{\iota j}, d_{\iota j}]},\, j \in I_{\iota 1} \end{array} \right\}, \qquad (10.3)$$

where $r_\iota = \max\{m_\iota, n_\iota\}$, and $C^{r_\iota}(J, \mathbb{R})$ are Banach spaces, equipped with the norms $\| \cdot \|_{C^{r_\iota}}$, defined by

$$\|w\|_{C^{r_\iota}} := \max\left\{ \|w^{(i)}\|_\infty,\, i \in I_\iota \right\}$$

and $\|y\|_\infty := \max\limits_{t \in [0, 1]} |y(t)|$. Analogously,

$$\|y\|_{[c_{\iota j}, d_{\iota j}]} = \max\limits_{t \in [c_{\iota j}, d_{\iota j}]} |y(t)|.$$

Furthermore, the product space $E := K_1 \times K_2$ with the norm

$$\|(u_1, u_2)\|_E := \max\left\{ \|u_1\|_{C^{r_1}}, \|u_2\|_{C^{r_2}} \right\}, \qquad (10.4)$$

is also a Banach space.

Remark 10.1. Assumption (A5) guarantees that K_ι are cones, for $\iota = 1, 2$.

10.2   Main result

The nonlinearities will verify either one of the following growth assumptions:

(B1) For $\iota = 1, 2$, $l = 0, 1, \ldots, m_\iota$, $j = 0, 1, \ldots, n_\iota$,

$$\limsup_{|x_l| \to 0, \, |y_j| \to 0} \max_{t \in J} \frac{f_\iota(t, x_0, \ldots, x_{m_\iota}, y_0, \ldots, y_{n_\iota})}{\max\{|x_l|, |y_j|\}} = 0 \qquad (10.5)$$

and

$$\liminf_{|x_l| \to +\infty, \, |y_j| \to +\infty} \min_{t \in J} \frac{f_\iota(t, x_0, \ldots, x_{m_\iota}, y_0, \ldots, y_{n_\iota})}{\max\{|x_l|, |y_j|\}} = +\infty; \qquad (10.6)$$

(B2) For $\iota = 1, 2$, $l = 0, 1, \ldots, m_\iota$, $j = 0, 1, \ldots, n_\iota$,

$$\liminf_{|x_l| \to 0, \, |y_j| \to 0} \min_{t \in J} \frac{f_\iota(t, x_0, \ldots, x_{m_\iota}, y_0, \ldots, y_{n_\iota})}{\max\{|x_l|, |y_j|\}} = +\infty \qquad (10.7)$$

and

$$\limsup_{|x_l| \to +\infty, \, |y_j| \to +\infty} \max_{t \in J} \frac{f_\iota(t, x_0, \ldots, x_{m_\iota}, y_0, \ldots, y_{n_\iota})}{\max\{|x_l|, |y_j|\}} = 0. \qquad (10.8)$$

The main result is given by the next theorem and provides the existence of at least one nontrivial solution of problem (10.2).

Theorem 10.2. *Let, for $\iota = 1, 2$, $f_\iota : J \times \mathbb{R}^{m_\iota + n_\iota + 2} \to [0, \infty)$ be L^∞-Carathéodory functions such that assumptions $(A1)$-$(A5)$ hold and or condition $(B1)$, or $(B2)$, is satisfied. Then problem (10.2) has at least one nontrivial solution $(u, v) \in \left(C^{r_1}(J) \times C^{r_2}(J) \right)$.*

Proof. Consider the cones K_ι in (10.3), and the Banach space $E := K_1 \times K_2$ with the norm given by (10.4). Define the operators $T_1 : E \to K_1$, $T_2 : E \to K_2$ and $T : E \to E$ given by

$$T_1(u_1, u_2)(t) = \int_0^1 k_1(t, s) \, g_1(s) \, f_1 \left(\begin{array}{c} s, u_1(s), \ldots, u_1^{(m_1)}(s), \\ u_2(s), \ldots, u_2^{(n_1)}(s) \end{array} \right) ds$$

$$T_2(u_1, u_2)(t) = \int_0^1 k_2(t, s) \, g_2(s) \, f_2 \left(\begin{array}{c} s, u_1(s), \ldots, u_1^{(m_2)}(s), \\ u_2(s), \ldots, u_2^{(n_2)}(s) \end{array} \right) ds \qquad (10.9)$$

and $T = (T_1, T_2)$.

The proof of the main theorem will be done in several steps and made only for the operator T_1, since the procedure for the operator T_2 is analogous. For this purpose, we need to show that T verifies the assumptions of Theorem 10.1, and consequently has a fixed point which is the solution of (10.2).

Step 1: $T : E \to E$ is *well defined in* E.

To prove that T_ι are well defined on K_ι, for $\iota = 1, 2$, take $(u_1, u_2) \in E$. By (A3) and Definition 9.1, for $i = 0, 1, \ldots, r_1$,

$$\| (T_1 (u_1, u_2))^{(i)} \|$$

$$= \max_{t \in [0, 1]} \left| \int_0^1 \frac{\partial^i k_1}{\partial t^i} (t, s) g_1(s) \, f_1(s, u_1(s), \ldots, u_1^{(m_1)}(s), u_2(s), \ldots, u_2^{(n_1)}(s)) \, ds \right|$$

$$\leq \int_0^1 \phi_{1i}(s) g_1(s) \, f_1(s, u_1(s), \ldots, u_1^{(m_1)}(s), u_2(s), \ldots, u_2^{(n_1)}(s)) \, ds$$

$$\leq \int_0^1 \phi_{1i}(s) \, g_1(s) \, \varphi_{\|(u_1, u_2)\|_E} \, ds < \infty,$$

which implies that $(T_1 (u_1, u_2))^{(i)} \in C^{r_1} (J, \mathbb{R})$.

By (A2), for $i \in I_{10}$, $(T_1 (u_1, u_2))^{(i)} (t) \geq 0$ for $t \in [q_{1i}, p_{1i}]$.

By (A4), for $j \in I_{11}$ and $t \in [c_{1j}, d_{1j}]$,

$$\left| (T_1 (u_1, u_2))^{(j)} (t) \right|$$

$$\leq \int_0^1 \left| \frac{\partial^j k_1}{\partial t^j} (t, s) \right| g_1(s) \, f_1(s, u_1(s), \ldots, u_1^{(m_1)}(s), u_2(s), \ldots, u_2^{(n_1)}(s)) \, ds$$

$$\leq \int_0^1 \psi_{1j}(s) \, g_1(s) \, f_1(s, u_1(s), \ldots, u_1^{(m_1)}(s), u_2(s), \ldots, u_2^{(n_1)}(s)) \, ds.$$

Applying the maximum for $t \in [c_{1j}, d_{1j}]$, we obtain

$$\| (T_1 (u_1, u_2))^{(j)} \|_{[c_{1j}, d_{1j}]}$$

$$\leq \int_0^1 \psi_{1j}(s) \, g_1(s) \, f_1(s, u_1(s), \ldots, u_1^{(m_1)}(s), u_2(s), \ldots, u_2^{(n_1)}(s)) \, ds.$$

Therefore, for $t \in [a_{1j}, b_{1j}]$, one has, for $j \in I_{11}$,

$$(T_1 (u_1, u_2))^{(j)} (t)$$

$$= \int_0^1 \frac{\partial^j k_1}{\partial t^j} (t, s) g_1(s) \, f_1(s, u_1(s), \ldots, u_1^{(m_1)}(s), u_2(s), \ldots, u_2^{(n_1)}(s)) \, ds$$

$$\geq \int_0^1 \xi_{1j} \psi_{1j} \, g_1(s) \, f_1(s, u_1(s), \ldots, u_1^{(m_1)}(s), u_2(s), \ldots, u_2^{(n_1)}(s)) \, ds$$

$$\geq \| (T_1 (u_1, u_2))^{(j)} \|_{[c_{1j}, d_{1j}]},$$

and so,

$$\min_{t \in [a_{1j}, b_{1j}]} (T_1 (u_1, u_2))^{(j)} (t) \geq \xi_{1j} \| (T_1 (u_1, u_2))^{(i)} \|.$$

Therefore, $T_1 E \subseteq K_1$. The inclusion $T_2 E \subseteq K_2$ can be proved similarly, and consequently $TE \subset E$.

Step 2: *T is uniformly bounded in E.*

We will prove that T_1 is uniformly bounded in K_1. The arguments to prove that T_2 is uniformly bounded in K_2, are analogous.

Consider

$$B := \{(u_1, u_2) \in E : \| (u_1, u_2) \|_E \leq \rho\}$$

for some $\rho > 0$.

By (A3) and Definition 9.1, for $j \in I_{11}$,

$$\| (T_1 (u_1, u_2))^{(j)} \|_\infty$$

$$= \max_{t \in J} \left| \int_0^1 \frac{\partial^{(j)} k_1}{\partial t^j} (t, s) g_1(s) \, f_1(s, u_1(s), \ldots, u_1^{(m_1)}(s), u_2(s), \ldots, u_2^{(n_1)}(s)) \, ds \right|$$

$$\leq \int_0^1 \phi_{1j}(s) g_1(s) \, f_1(s, u_1(s), \ldots, u_1^{(m_1)}(s), u_2(s), \ldots, u_2^{(n_1)}(s)) \, ds$$

$$\leq \int_0^1 \phi_{1j}(s) g_1(s) \, \varphi_\rho(s) ds := M_{1j} > 0.$$

Therefore

$$\|T_1 (u_1, u_2) \|_{C^{r_1}} < \max \{M_{1j} : j \in I_{11}\},$$

for all $(u_1, u_2) \in E$ such that $\| (u_1, u_2) \|_E \leq \rho$.

So, $T_1(B)$ is uniformly bounded in K_1. Analogously $T_2(B)$ is uniformly bounded in K_2, and, then, $T(B)$ is uniformly bounded in E.

Step 3: *T(B) is equicontinuous in E.*

This step will be shown if $T_1(B)$ and $T_2(B)$ are equicontinuous in K_1 and K_2, respectively. The calculus will be done only for $T_1(B)$, as the other case is similar.

Consider $t_1, t_2 \in [0, 1]$. By (A1), for $i = 0, 1, \ldots, r_1 - 1$,

$$|(T_1 (u_1, u_2))^{(i)}(t_1) - (T_1 (u_1, u_2))^{(i)}(t_2)|$$

$$\leq \int_0^1 \left| \frac{\partial^i k_1}{\partial t^i}(t_1, s) - \frac{\partial^i k_1}{\partial t^i}(t_2, s) \right| g_1(s) f_1 \left(\begin{array}{c} s, u_1(s), \ldots, u_1^{(m_1)}(s), \\ u_2(s), \ldots, u_2^{(n_1)}(s) \end{array} \right) ds$$

$$\leq \int_0^1 \left| \frac{\partial^i k_1}{\partial t^i}(t_1, s) - \frac{\partial^i k_1}{\partial t^i}(t_2, s) \right| g_1(s) \, \varphi_\rho(s) ds \to 0, \text{ as } t_1 \to t_2.$$

Without loss of generality suppose $t_1 \leq t_2$. Using the same arguments as above, for the r_1-*th* derivative, we have

$$\left| (T_1\,(u_1, u_2))^{(r_1)}(t_1) - (T_1\,(u_1, u_2))^{(r_1)}(t_2) \right|$$

$$\leq \int_0^1 \left| \frac{\partial^{r_1} k_1}{\partial t^{r_1}}(t_1, s) - \frac{\partial^{r_1} k_1}{\partial t^{r_1}}(t_2, s) \right| g_1(s)\, f_1 \left(\begin{array}{c} s, u_1(s), \ldots, u_1^{(m_1)}(s), \\ u_2(s), \ldots, u_2^{(n_1)}(s) \end{array} \right) ds$$

$$\leq \int_0^1 \left| \frac{\partial^{r_1} k_1}{\partial t^{r_1}}(t_1, s) - \frac{\partial^{r_1} k_1}{\partial t^{r_1}}(t_2, s) \right| g_1(s)\, \varphi_\rho(s) ds$$

$$= \int_0^{t_1} \left| \frac{\partial^{r_1} k_1}{\partial t^{r_1}}(t_1, s) - \frac{\partial^{r_1} k_1}{\partial t^{r_1}}(t_2, s) \right| g_1(s)\, \varphi_\rho(s) ds$$

$$+ \int_{t_1}^{t_2} \left| \frac{\partial^{r_1} k_1}{\partial t^{r_1}}(t_1, s) - \frac{\partial^{r_1} k_1}{\partial t^{r_1}}(t_2, s) \right| g_1(s)\, \varphi_\rho(s) ds$$

$$+ \int_{t_2}^1 \left| \frac{\partial^{r_1} k_1}{\partial t^{r_1}}(t_1, s) - \frac{\partial^{r_1} k_1}{\partial t^{r_1}}(t_2, s) \right| g_1(s)\, \varphi_\rho(s) ds \to 0, \ \text{as} \ t_1 \to t_2.$$

Therefore, $T_1(B)$ is equicontinuous in K_1.

By the Arzelà-Ascoli Theorem, T is completely continuous in E.

Suppose that condition (B1) is verified.

Step 4: $T\,(u_1, u_2) \not\geq (u_1, u_2)$, for all $(u_1, u_2) \in E \cap \partial\Omega_1$ with $\Omega_1 = \{(u_1, u_2) \in E : \|(u_1, u_2)\|_E < \rho_1\}$, for some $\rho_1 > 0$.

To prove this step it will be enough to show that, for some $(u_1, u_2) \in E \cap \partial\Omega_1$,

$$\|T_1\,(u_1, u_2)\|_{C^{r_1}} \leq \|(u_1, u_2)\|_E \ \text{or} \ \|T_2\,(u_1, u_2)\|_{C^{r_2}} \leq \|(u_1, u_2)\|_E.$$

Consider $(u_1, u_2) \in E \cap \partial\Omega_1$. Therefore $\|(u_1, u_2)\|_E = \rho_1$.
For $j \in I_{11}$, and (A4), define

$$\varepsilon_1 := \min \left\{ \frac{1}{\int_0^1 \phi_{1i}(s) g_1(s)\, ds} \right\}. \tag{10.10}$$

By (9.5), there exists $0 < \rho_1 < 1$ such that

$$f_1(t, u_1(t), \ldots, u_1^{(m_1)}(t), u_2(t), \ldots, u_2^{(n_1)}(t)) \leq \varepsilon_1 \|(u_1, u_2)\|_E, \tag{10.11}$$

for $\|(u_1, u_2)\|_E \leq \rho_1$.
By (A3), (10.11), and (10.10), for $i = 0, 1, \ldots, r_1$,

$$(T_1\,(u_1, u_2)\,(t))^{(i)} \leq \int_0^1 \left| \frac{\partial^i k_1}{\partial t^i}(t, s) \right| g_1(s)\, f_1 \left(\begin{array}{c} s, u_1(s), \ldots, u_1^{(m_1)}(s), \\ u_2(s), \ldots, u_2^{(n_1)}(s) \end{array} \right) ds$$

$$\leq \int_0^1 \phi_{1i}(s) g_1(s)\, \varepsilon_1 \|(u_1, u_2)\|_E\, ds < \|(u_1, u_2)\|_E = \rho_1.$$

Therefore, $\|T_1(u_1, u_2)\|_{C^{r_1}} \leq \|(u_1, u_2)\|_E$, for $(u_1, u_2) \in E \cap \partial\Omega_1$.

By similar arguments it can be shown that $\|T_2(u_1, u_2)\|_{C^{r_2}} \leq \|(u_1, u_2)\|_E$ and therefore, $\|T(u_1, u_2)\|_E \leq \|(u_1, u_2)\|_E$, for $(u_1, u_2) \in E \cap \partial\Omega_1$.

Step 5: $T(u_1, u_2) \not\leq (u_1, u_2)$, *for all* $u \in E \cap \partial\Omega_2$ *with* $\Omega_2 = \{(u_1, u_2) \in E : \|(u_1, u_2)\|_E < \rho_2\}$, *for some* $\rho_2 > 0$.

By (9.6), for $\iota = 1, 2$, there exist $\rho_\iota^* > 0$ and $\theta > 0$, such that, when $\|(u_1, u_2)\|_E \geq \theta$ we have

$$f_1(t, u_1(t), \dots, u_1^{(m_1)}(t), u_2(t), \dots, u_2^{(n_1)}(t)) \geq \|(u_1, u_2)\|_E. \tag{10.12}$$

Define, for $j \in I_{11}$,

$$\eta_1 := \max\left\{ \frac{1}{\xi_{1j} \int_{a_{1j}}^{b_{1j}} \psi_{1j}(s) g_1(s)\, ds} \right\}. \tag{10.13}$$

Let $(u_1, u_2) \in E$ be such that $\|(u_1, u_2)\|_E = \rho_2$, with $\rho_2 > \rho_1$.

Now from (A4), and (9.13), for $j \in I_{11}$, and $t \in [a_{1j}, b_{1j}]$

$$\begin{aligned}
(T_1(u_1, u_2)(t))^{(j)} &= \int_0^1 \frac{\partial^{(j)} k_1}{\partial t^j}(t, s) g_1(s) f_1\left(\begin{array}{l} s, u_1(s), \dots, u_1^{(m_1)}(s), \\ u_2(s), \dots, u_2^{(n_1)}(s) \end{array} \right) ds \\
&\geq \int_{a_{1j}}^{b_{1j}} \frac{\partial^{(j)} k_1}{\partial t^j}(t, s) g_1(s) f_1\left(\begin{array}{l} s, u_1(s), \dots, u_1^{(m_1)}(s), \\ u_2(s), \dots, u_2^{(n_1)}(s) \end{array} \right) ds \\
&\geq \xi_{1j} \int_{a_{1j}}^{b_{1j}} \psi_{1j}(s) g_1(s) f_1\left(\begin{array}{l} s, u_1(s), \dots, u_1^{(m_1)}(s), \\ u_2(s), \dots, u_2^{(n_1)}(s) \end{array} \right) ds \\
&\geq \xi_{1j} \eta_1 \rho_2 \int_{a_{1j}}^{b_{1j}} \psi_{1j}(s) g_1(s)\, ds \geq \rho_2 = \|(u_1, u_2)\|_E \\
&\geq (u_1, u_2)(t).
\end{aligned}$$

Therefore, $T_1(u_1, u_2) \not\leq (u_1, u_2)$ for all $(u_1, u_2) \in E \cap \partial\Omega_2$, and, so, $T(u_1, u_2) \not\leq (u_1, u_2)$ for all $(u_1, u_2) \in E \cap \partial\Omega_2$.

By Theorem 10.1, (i), the operator T has a fixed point in $K \cap (\overline{\Omega_2} \setminus \Omega_1)$ which in turn is a solution of our problem.

Now assume that (B2) holds.

Step 6: $T_1(u_1, u_2) \not\leq (u_1, u_2)$ for all $(u_1, u_2) \in E \cap \partial\Omega_3$, with $\Omega_3 = \{(u_1, u_2) \in E : \|(u_1, u_2)\|_E < \rho_3\}$, for some $\rho_3 > 0$.

Taking $\eta_1 > 0$ as in (10.13), we see that, by (10.7), there exists $0 < \rho_3 < 1$ such that

$$(t, u_1(t), \ldots, u_1^{(m_1)}(t), u_2(t), \ldots, u_2^{(n_1)}(t)) \in [0, 1] \times [0, \rho_{3_,}]^{m_1+n_1+2}$$

and

$$f_1(t, u_1(t), \ldots, u_1^{(m_1)}(t), u_2(t), \ldots, u_2^{(n_1)}(t)) \geq \eta_1 \| (u_1, u_2) \|_E. \qquad (10.14)$$

Consider $(u_1, u_2) \in E$ such that $\| (u_1, u_2) \|_E = \rho_3$. Then, applying similar inequalities as in Step 5, we obtain that $T_1(u_1, u_2) \nleq (u_1, u_2)$ for all $(u_1, u_2) \in E \cap \partial\Omega_3$.

So, $T(u_1, u_2) \nleq (u_1, u_2)$ for all $(u_1, u_2) \in E \cap \partial\Omega_3$.

Step 7: $T(u_1, u_2) \ngeq (u_1, u_2)$, for all $(u_1, u_2) \in E \cap \partial\Omega_4$ with $\Omega_4 = \{(u_1, u_2) \in E : \| (u_1, u_2) \|_E < \rho_4\}$, for some $\rho_4 > 0$.

Case 7.1. *Suppose that f_1 is bounded.*

Then there is $N > 0$ such that $f_1(t, u_1(t), \ldots, u_1^{(m_1)}(t), u_2(t), \ldots, u_2^{(n_1)}(t)) \leq N$ for all $(t, u_1(t), \ldots, u_1^{(m_1)}(t), u_2(t), \ldots, u_2^{(n_1)}(t)) \in [0, 1] \times [0, +\infty)^{m_1+n_1+2}$. Choose

$$\rho_4 := \max \left\{ \rho_3 + 1, N \int_0^1 \phi_{1i}(s) g_1(s) \, ds : i = 0, 1, \ldots, r_1 \right\}$$

and take $(u_1, u_2) \in E$ with $\| (u_1, u_2) \|_E = \rho_4$. Then, for $i = 0, 1, \ldots, r_1$,

$$(T_1(u_1, u_2)(t))^{(i)}$$
$$\leq \int_0^1 \left| \frac{\partial^i k_1}{\partial t^i}(t, s) \right| g_1(s) \, f_1(s, u_1(s), \ldots, u_1^{(m_1)}(s), u_2(s), \ldots, u_2^{(n_1)}(s)) ds$$
$$\leq N \int_0^1 \phi_{1i}(s) g_1(s) \, ds \leq \rho_4, \text{ for } t \in [0, 1].$$

Thus, $\|T_1(u_1, u_2)\|_{C^{r_1}} \leq \| (u_1, u_2) \|_E$, and, therefore, $T(u_1, u_2) \ngeq (u_1, u_2)$, for all $(u_1, u_2) \in E \cap \partial\Omega_4$.

Case 7.2. *Suppose that f_1 is unbounded.*

By (10.8), there exists $\mu > 0$ such that

$$\max \left\{ \mu \int_0^1 \phi_{1i}(s) g_1(s) ds : i = 0, 1, \ldots, r_1 \right\} \leq 1 \qquad (10.15)$$

and

$$f_1(t, u_1(t), \ldots, u_1^{(m_1)}(t), u_2(t), \ldots, u_2^{(n_1)}(t)) \leq \mu \| (u_1, u_2) \|_E, \qquad (10.16)$$

for every $M > 0$ such that $\| (u_1, u_2) \|_E \geq M$.

Define
$$\rho_4 := \max\{M, \rho_3 + 1\}.$$
Then, for $(u_1, u_2) \in E \cap \partial\Omega_4$ we have $\|(u_1, u_2)\|_E = \rho_4$ and, by (10.16),
$$f_1(t, u_1(t), \dots, u_1^{(m_1)}(t), u_2(t), \dots, u_2^{(n_1)}(t)) \leq \mu\|(u_1, u_2)\|_E = \mu\rho_4.$$
$$(10.17)$$

So, for $i = 0, 1, \dots, r_1$,
$$(T_1(u_1, u_2)(t))^{(i)}$$
$$\leq \int_0^1 \left|\frac{\partial^i k_1}{\partial t^i}(t, s)\right| g_1(s) \, f_1(s, u_1(s), \dots, u_1^{(m_1)}(s), u_2(s), \dots, u_2^{(n_1)}(s)) ds$$
$$\leq \int_0^1 \phi_{1i}(s) g_1(s) \, \mu\rho_4 \, ds \leq \mu\rho_4 \int_0^1 \phi_{1i}(s) g_1(s) \, ds \leq \rho_4.$$
Therefore, $\|T_1(u_1, u_2)\|_{C^{r_1}} \leq \|(u_1, u_2)\|_E$, for $(u_1, u_2) \in E \cap \partial\Omega_4$, and $T(u_1, u_2) \not\geq (u_1, u_2)$, for all $(u_1, u_2) \in E \cap \partial\Omega_4$.

The remaining cases (f_2 bounded or unbounded) can be studied by similar techniques.

By Theorem 10.1, the operator T has a fixed point in $E \cap (\overline{\Omega_4} \backslash \Omega_3)$ that, in turn, is a solution of the problem (10.2). $\qquad \square$

10.3 Coupled system for a suspension bridge bending model

Beams are narrow elements that support loads applied perpendicular to their longitudinal axis, that is, they are bar-shaped structures whose main purpose is to support transverse loads and transport them to the supports. For example, elements to support the floor of a building, the deck of a bridge, the wing of an airplane, or the axis of an automobile. In fact, beams are important structural and mechanical elements used in engineering projects, particularly in Civil and Mechanical Engineering (see [97]).

The suspension bridge is viewed as an elastic beam which is suspended to a sustaining cable, where the beam and the cable are connected by a large number of hangers (see [187,264]). The system described is illustrated in Figure 10.1 and can be modeled by a fourth-order ordinary differential problem and hinged boundary conditions
$$\begin{cases} EIw^{(4)}(x) - H\left(\frac{w'(x)}{1+(y'(x))^2}\right) - \frac{w''(x)+y''(x)}{(1+(w'(x)+y'(x))^2)^{3/2}} h(w) = p, \quad x \in (0, L) \\ \\ w(0) = w(L) = w''(0) = w''(L) = 0, \end{cases}$$
$$(10.18)$$

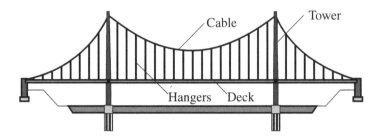

Fig. 10.1   Beam sustained by a cable through parallel hangers

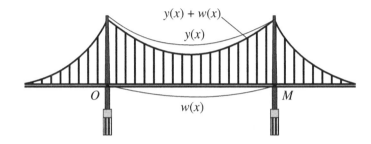

Fig. 10.2   Beam sustained by a cable through parallel hangers with additional tension

where

- $w = w(x)$ is the vertical displacement of the beam;
- $y = y(x)$ is the position of the cable at rest;
- L is the length of the roadway between the two towers;
- E and I are the elastic modulus of the material and the moment of inertia of the cross section, respectively;
- H is the horizontal tension in the cable subject to the dead load $q = q(x)$;
- $h(w)$ is the additional tension in the cable produced by the live load $p = p(x)$.

As illustrated in Figure 10.2, the system (the suspension bridge) is subject to an action of dead load $q(x)$, including the weight of the cable, the weight of the hangers and the dead weight of the roadway without producing a bending moment in the beam. At this moment, the cable is in the position $y(x)$, while the unloaded beam is the segment connecting O (origin of the orthogonal coordinate system and positive displacements are oriented downwards) and M (has the coordinate $(0, L)$, where L is the length of the roadway between two towers). Then the horizontal component $H > 0$ of the tension remains constant. Therefore, there is an equilibrium position in the system given by the equation

$$Hy''(x) = -q(x). \tag{10.19}$$

Applying the arguments in the references above, we consider the equation (10.18) combined by (10.19), together with some boundary conditions in order to obtain the following normalized coupled system ($L = 1$) composed by the fourth and second order nonlinear equations which includes the effect of the shear force ($w'''(x)$) (see [249]):

$$\begin{cases} EIw^{(4)}(x) - H\frac{|w'(x)|}{1+(y'(x))^2} - \frac{w''(x)+y''(x)}{(1+(w'(x)+y'(x))^2)^{3/2}} \\ \qquad h(w) = p(x)\left(w'''(x)\right)^2, x \in (0,1), \\ \\ \qquad -y''(x) = \frac{q(x)}{H}\left(w'''(x)\right)^3 \\ \\ \qquad w(0) = w(1) = w''(0) = w''(1) = 0 \\ \\ \qquad y(0) = 1, \ y(1) = 0, \end{cases} \tag{10.20}$$

where H, h, p, q are non negative forces.

The problem (10.20) is a particular case of (10.2), since it can be rewritten as the following system of integral equations

$$\begin{cases} w(x) = \int_0^1 k_1(x,s) \frac{1}{EI} \left[\begin{array}{l} p(s)\left(w'''(s)\right)^2 + H\frac{|w'(s)|}{1+(y'(s))^2} \\ + \frac{w''(s)+y''(s)}{(1+(w'(s)+y'(s))^2)^{3/2}}h(w(s)) \end{array} \right] ds, \\ \\ y(x) = 1 + \int_0^1 k_2(x,s) \frac{q(s)}{H}\left(w'''(s)\right)^3 ds, \end{cases} \tag{10.21}$$

where the kernel functions $k_1(x,s)$ and $k_2(x,s)$ are given, respectively, by the corresponding Green's functions

$$G_1(x,s) = \frac{1}{6} \begin{cases} x(1-s)(2s-s^2-x^2), & 0 \le x \le s \le 1, \\ s(1-x)(2x-x^2-s^2), & 0 \le s \le x \le 1 \end{cases}$$

$$G_2(x,s) = \begin{cases} x(1-s), & 0 \le x \le s \le 1, \\ s(1-x), & 0 \le s \le x \le 1 \end{cases}$$

with the first and second derivatives of G_1 given by

$$\frac{\partial G_1}{\partial x}(x,s) = \frac{1}{6}\begin{cases}(1-s)(2s-s^2-3x^2), & 0 \le x \le s \le 1,\\ s(3x^2-6x+s^2+2), & 0 \le s \le x \le 1,\end{cases}$$

$$\frac{\partial^2 G_1}{\partial x^2}(x,s) = \begin{cases}x(s-1), & 0 \le x \le s \le 1,\\ s(x-1), & 0 \le s \le x \le 1,\end{cases}$$

$r_1 = m_1 = 3, n_1 = 2, r_2 = m_2 = 3, g_1(x) = \frac{1}{EI}, g_2(x) = 1$, with $EI > 0$, $q(x) > 0$ and

$$f_1(x,y_1,\ldots,y_7) = p(x)(y_4)^2 + H\frac{y_2}{1+y_6^2} + \frac{y_3+y_7}{(1+(y_2+y_6)^2)^{3/2}}h(y_1),$$
$$\tag{10.22}$$

$$f_2(x,y_4) = \frac{q(x)}{H}(y_4)^3. \tag{10.23}$$

Note that condition (A1) holds trivially.

As $(x,s) \in [0,1] \times [0,1]$, then the Green's functions G_1 and G_2 are positive on $[0,1] \times [0,1]$.

On the other hand, G_1 is positive on $(x_1,s_1) \in \left[0, \frac{2}{3+\sqrt{3}}\right] \times \left[0, \frac{2}{3+\sqrt{3}}\right]$ and $\frac{\partial^2 G_1}{\partial x^2}(x,s) \ge 0$ on the boundary of the square, that is, on

$$\{(0,s),(1,s), s \in [0,1]\} \cup \{(x,0),(x,1), x \in [0,1]\}.$$

Therefore, (A2) holds. The assumption (A5) is also satisfied, because $[q_{10}, p_{10}] \equiv J$ and $[q_{20}, p_{20}] \equiv J$, as $G_1(x,s)$ and $G_1(x,s)$ are positive on $[0,1] \times [0,1]$.

In order to show (A3) holds, we have

$$|G_1(x,s)| \le \frac{1}{6}s(2-s^2) := \phi_{10}, \text{ for } x \in [0,1], \ s \in [0,1],$$

$$\left|\frac{\partial G_1}{\partial x}(x,s)\right| \le \frac{1}{6}s(2+s^2) := \phi_{11}, \text{ for } x \in [0,1], \ s \in [0,1],$$

$$\left|\frac{\partial^2 G_1}{\partial x^2}(x,s)\right| \le s(1-s) := \phi_{12}, \text{ for } x \in [0,1], \ s \in [0,1],$$

and

$$|G_2(x,s)| \le (1-s) := \phi_{20}, \text{ for } x \in [0,1], \ s \in [0,1].$$

Finally to prove that (A4) is satisfied, take

$$\xi_{10} = \frac{1}{m_1}, \ \xi_{11} = \frac{1}{m_2}, \ \xi_{12} = \frac{1}{m_3}, \ \xi_{20} = \frac{1}{m_4},$$

where m_i, $i = 1, \ldots, 4$ are positive numbers such that $\xi_{10}, \xi_{11}, \xi_{12}, \xi_{20} \in (0, 1)$. Then (A4) is verified, by simply taking $m_i > 1$, $i = 1, \ldots, 4$, large enough such that,

$$G_1(x, s) \geq \frac{1}{m_1 6} s(2 - s^2) := \xi_{10}\phi_{10}, \quad \frac{\partial G_1}{\partial x}(x, s) \geq \frac{1}{m_2 6} s(2 + s^2) := \xi_{11}\phi_{11},$$

$$\frac{\partial^2 G_1}{\partial x^2}(x, s) \geq \frac{1}{m_3} s(1 - s) := \xi_{12}\phi_{12}, \quad G_2(x, s) \geq \frac{1}{m_4} s(1 - s) := \xi_{20}\phi_{20},$$

for $x \in [0, 1]$ and $s \in [0, 1]$.

The functions $f_1 : [0, 1] \times [0, \infty)^7 \to [0, \infty)$, $f_2 : [0, 1] \times [0, \infty) \to [0, \infty)$, given by (10.22) and (10.23), respectively, are L^∞−Carathéodory functions, as, for $\rho > 0$ such that

$$\max\{|y_1|, |y_2|, |y_3|, |y_4|, |y_5|, |y_6|, |y_7|\} < \rho,$$

there exist functions $\varphi_{1\rho}, \varphi_{2\rho} \in L^\infty(J)$ verifying

$$f_1(x, y_1, y_2, y_3, y_4, y_5, y_6, y_7)$$
$$\leq \max_{x \in J} p(x)\rho^2 + H\frac{\rho}{1 + \rho^2} + \frac{2\rho}{(1 + 4\rho^2)^{3/2}} h(\rho) := \varphi_{1\rho}$$

$$f_2(x, y_4) \leq \frac{\max_{x \in J} q(x)}{H}\rho^3 := \varphi_{2\rho}.$$

For $j = 0, 1, 2, 3$ and $l = 0, 1, 2$, we have

$$\limsup_{w^{(j)}, y^{(l)} \to 0} \max_{x \in [0,1]} \frac{\frac{1}{EI}\left[p(x)(w'''(x))^2 + H\frac{|w'(x)|}{1+(y'(x))^2} + \frac{w''(x)+y''(x)}{(1+(w'(x)+y'(x))^2)^{3/2}} h(w(x)) \right]}{\max\{|w^{(j)}|, |y^{(l)}|\}} = 0,$$

$$\limsup_{w^{(j)}, y^{(l)} \to +\infty} \max_{x \in [0,1]} \frac{\frac{q(x)}{H}(w'''(x))^3}{\max\{|w^{(j)}|, |y^{(l)}|\}} = 0,$$

and

$$\liminf_{w^{(j)}, y^{(l)} \to +\infty} \min_{x \in [0,1]} \frac{\frac{1}{EI}\left[p(x)(w'''(x))^2 + H\frac{|w'(x)|}{1+(y'(x))^2} + \frac{w''(x)+y''(x)}{(1+(w'(x)+y'(x))^2)^{3/2}} h(w(x)) \right]}{\max\{|w^{(j)}|, |y^{(l)}|\}} = +\infty,$$

$$\liminf_{w^{(j)}, y^{(l)} \to +\infty} \min_{x \in [0,1]} \frac{\frac{q(x)}{H}(w'''(x))^3}{\max\{|w^{(j)}|, |y^{(l)}|\}} = +\infty,$$

and, therefore, condition (B1) holds.

So, by Theorem 10.2, there is at least one positive solution $(u, v) \in (C^3[0, 1] \times C^3[0, 1])$ of problem (10.21), which is the solution of (10.20).

Chapter 11

Generalized Hammerstein coupled systems on the real line

Integral equations are of many types and Hammerstein-type is a particular case of them. These equations appear naturally in inverse problems, fluid dynamics, potential theory, spread of interdependent epidemics, elasticity, etc. (see [4, 7, 106]). Hammerstein-type integral equations usually arise from the reformulation of boundary value problems associated to partial or ordinary differential equations.

Hammerstein-type integral equations in real line play an important role in physical problems and are often used to reformulate or rewrite mathematical problems. For example, the propagation of mono-frequency acoustic or electromagnetic waves over flat nonhomogeneous terrain modeled by the Helmholtz equation

$$\triangle\varphi + k^2\varphi = 0$$

in the upper half plane $D = \{(x_1, x_2) \subset \mathbb{R}^2 : x_2 > 0\}$ with a Robin or impedance condition

$$\frac{\partial\triangle}{\partial x_2} + ik\beta\varphi = \varphi_0$$

on the boundary line ∂D, where k is a wave number, $\beta \in L^\infty(\partial D)$ is the surface admittance describing the local properties of the ground surface ∂D and $\varphi_0 \in L^\infty(\partial D)$ the inhomogeneous term, can be reformulated as Hammerstein-type integral equations (see [60]). In fact, the authors have shown that the above problem is equivalent to Hammerstein-type integral equations in the real line

$$u(x) - \int_{-\infty}^{+\infty} \tilde{k}(x - y)z(y)u(y)dy = \psi(x), \ x \in \mathbb{R},$$

where ψ is given, $\tilde{k} \in L^1 \cap C(\mathbb{R} \setminus \{0\})$, $z \in L^\infty$ is closely connected with the surface admittance β ($z = i(1-\beta)$) and $u \in BC$ is to be determined. [78] introduced new variants of some nonlinear alternatives of Leray–Schauder and

Krasnosel'skij type involving the weak topology of Banach spaces. Along with the proof of theorems on the existence of solutions, profound constructive solvability theorems were proposed with analysis of branching solutions of nonlinear Hammerstein integral equation presented in [238]. Interested readers can find explicit and implicit parameterizations in the construction of branching solutions by iterative methods in [239].

However, due to the lack of compactness, there are only few studies in the literature on Hammerstein integral equations in the real line compared to works in bounded domains.

By reviewing existence results for various types of functional, differential, and integral equations, in [35] Banaś and Sadarangani use arguments associated with the measure of non-compactness and illustrate applications in proving existence for some integro-functional equations in the set of continuous functions.

Information on the utility, and some mathematical tools used to address Hammerstein-type integral equations in real line or half-line can be found in [5, 36, 56, 201, 215].

We also highlight recent works, not necessarily in real line or half-line, on Hammerstein-type integral equations, with several approaches and applications on [3, 66, 95, 100, 131, 180, 198, 200, 241, 272] and the references therein.

On the other hand, Cabada *et al.* [55] deal about Hammerstein-type integral equations in unbounded domains via spectral theory. More concretely they study the existence of fixed points of the integral operator

$$Tu(t) := \int_{-\infty}^{+\infty} k(t,s)\eta(s)f(s,u(s))ds,$$

where $f : \mathbb{R}^2 \to [0, +\infty)$ satisfies a sort of L^∞-Carathéodory condition, $k : \mathbb{R}^2 \to \mathbb{R}$ is the kernel function and $\eta(t) \geq 0$ for a.e. $t \in \mathbb{R}$.

Ilhan and Ozdemir [132], study the nonlinear perturbed integral equation

$$x(t) = (T_1 x)(t) + (T_2 x)(t) \int_0^{+\infty} u(t, s, x(s))ds, \quad t \in \mathbb{R}^+,$$

where the functions $u(t, s, x)$ and the operators $T_i x$, $(i = 1, 2)$ are given, while $x = x(t)$ is an unknown function. Using the technique of a suitable measure of noncompactness, they prove an existence theorem for the mentioned system.

Based on several fundamental assumptions and some necessary and sufficient conditions under which the solution blows up in finite time, in [47], Brunner and Yang investigate the blow-up behaviors of solutions of Hammerstein-Volterra-type equations

$$u(t) = \phi(t) + \int_0^t k(t-s)G(s,u(s))ds,$$

where $f : [0,\infty) \to [0,\infty)$ and $G : [0,\infty) \times \mathbb{R} \to [0,\infty)$ are continuous functions, the kernel $k : (0,\infty) \to [0,\infty)$ is a locally integrable function and u is an unknown continuous solution.

Motivated by the works above, we consider the following generalized coupled systems of integral equations of Hammerstein-type

$$\begin{cases} u_1(t) = \int_{-\infty}^{+\infty} k_1(t,s)\, g_1(s)\, f_1(s,u_1(s),\ldots,u_1^{(m_1)}(s),u_2(s),\ldots,u_2^{(n_1)}(s))\, ds, \\[2mm] u_2(t) = \int_{-\infty}^{+\infty} k_2(t,s)\, g_2(s)\, f_2(s,u_1(s),\ldots,u_1^{(m_2)}(s),u_2(s),\ldots,u_2^{(n_2)}(s))\, ds \end{cases}$$

(11.1)

where $k_\iota : \mathbb{R}^2 \to \mathbb{R}$, $\iota = 1,2$, are the kernel functions such that $k_\iota \in W^{r_\iota,1}(\mathbb{R}^2)$, $r_\iota = \max\{m_\iota, n_\iota\}$, with $m_\iota, n_\iota \geq 0$, $g_\iota \in L^1(\mathbb{R})$ with $g_\iota(t) \geq 0$ for a.e. $t \in \mathbb{R}$ integrable, and $f_\iota : \mathbb{R}^{m_\iota+n_\iota+3} \to \mathbb{R}$ are L^∞-Carathéodory functions.

The main existence tool is Schauder's fixed point theorem in the space of bounded and continuous functions with bounded and continuous derivatives on $\mathbb{R}$, combined with the equiconvergence at $\pm\infty$ to recover the compactness of the correspondent operators. To the best of our knowledge, it is the first time where coupled Hammerstein-type integral equations in real line are considered with nonlinearities depending on several derivatives of both variables and, moreover, the derivatives can be of different order on each variable and each equation. On the other hand, we emphasize that the kernel functions can change sign and their derivatives in order for the first variable to be discontinuous.

11.1 Auxiliary results and assumptions

For $\iota = 1,2$, let $r_\iota = \max\{m_\iota, n_\iota\}$ and consider the Banach spaces defined by $E_\iota := BC^{r_\iota}(\mathbb{R})$ (space of bounded and continuous functions with bounded and continuous derivatives on $\mathbb{R}$, till order r_ι).

The spaces E_ι defined above are equipped with the norms $\|\cdot\|_{E_\iota}$, where

$$\|w\|_{E_\iota} := \max\left\{\|w^{(i)}\|_\infty,\, i = 0,1,\ldots,r_\iota\right\}$$

and

$$\|w\|_\infty := \sup_{t\in\mathbb{R}} |w(t)|.$$

Besides, $E := E_1 \times E_2$ with the norm

$$\|(u_1, u_2)\|_E := \max\left\{\|u_1\|_{E_1}, \|u_2\|_{E_2}\right\},$$

is also a Banach space.

For the reader's convenience we precise L^∞-Carathéodory functions given by Definition 9.1.

To guarantee a convenient criterion for compacity, we consider Theorem 3.1, with condition *(iii)* replaced by:

$$\left| x^{(i)}(t) - \lim_{t\to\pm\infty} x^{(i)}(t) \right| < \epsilon, \ \forall |t| > t_\epsilon, \ i = 0, 1, \ldots, r_\iota \text{ and } x \in M,$$

and the existence tool will be given by Schauder's fixed point theorem (Theorem 5.2).

In this chapter we consider the following assumptions:

(A1) For $\iota = 1, 2$, the function $k_\iota : \mathbb{R}^2 \to \mathbb{R}$, $k_\iota \in W^{r_\iota, 1}\left(\mathbb{R}^2\right)$, verify for all $\tau \in \mathbb{R}$,

$$\lim_{t\to\pm\infty} k_\iota(t, s) \in \mathbb{R}, \quad \lim_{t\to\pm\infty} \left| \frac{\partial^i k_\iota}{\partial t^i}(t, s) \right| \in \mathbb{R},$$

for $i = 1, \ldots, r_\iota$, $s \in \mathbb{R}$, and

$$\lim_{t\to\tau} \left| \frac{\partial^i k_\iota}{\partial t^i}(t, s) - \frac{\partial^i k_\iota}{\partial t^i}(\tau, s) \right| = 0, \text{ for a.e. } s \in \mathbb{R} \text{ and } i = 0, 1, \ldots, r_\iota.$$

(A2) For $\iota = 1, 2$, there are positive continuous functions $\phi_{\iota j}$, $j = 0, 1, \ldots, r_\iota$, such that

$$\left| \frac{\partial^j k_\iota}{\partial t^j}(t, s) \right| \leq \phi_{\iota j}(s) \text{ for } t \in \mathbb{R}, \text{ a.e. } s \in \mathbb{R}$$

and

$$\int_{-\infty}^{+\infty} \phi_{\iota j}(s) g_\iota(s) \varphi_{\iota \rho}(s) ds < +\infty \text{ for } j = 0, 1, \ldots, r_\iota,$$

with $\varphi_{\iota \rho}$ given by Definition 9.1.

11.2   Main theorem

This section is dedicated to the main result of this chapter, that is, its statement and its proof and provides the existence of solution of problem (11.1).

Theorem 11.1. *Let for $\iota = 1, 2$, $f_\iota : \mathbb{R}^{m_\iota+n_\iota+3} \to \mathbb{R}$ be L^∞-Carathéodory functions, such that, for some $t \in \mathbb{R}$, $f_\iota(t, 0, \ldots, 0) \neq 0$, and $g_\iota \in L^1(\mathbb{R})$ with $g_\iota(t) \geq 0$ for a.e. $t \in \mathbb{R}$.*
Consider that assumptions $(A1)$, $(A2)$ hold, and, moreover, assume that there is $R > 0$, such that, for $j = 0, 1, \ldots, r_\iota$,

$$R > \max \left\{ \int_{-\infty}^{+\infty} \phi_{\iota j}(s) g_\iota(s) \varphi_R(s) ds \right\}, \tag{11.2}$$

where $\varphi_R \in L^\infty(\mathbb{R})$, $y \in \mathbb{R}^{m_\iota+n_\iota+2}$ with $|f_\iota(t, y)| \leq \varphi_R(t)$, a.e. $t \in \mathbb{R}$. Then problem (11.1) has a nontrivial solution $(u, v) \in E_1 \times E_2$.

Proof. Consider the operators $T_1 : E \to E_1$ and $T_2 : E \to E_2$ such that

$$
\begin{cases}
T_1(u_1, u_2)(t) = \int_{-\infty}^{+\infty} k_1(t, s) \, g_1(s) \, f_1 \begin{pmatrix} s, u_1(s), \ldots, u_1^{(m_1)}(s), \\ u_2(s), \ldots, u_2^{(n_1)}(s) \end{pmatrix} ds \\[3mm]
T_2(u_1, u_2)(t) = \int_{-\infty}^{+\infty} k_2(t, s) \, g_2(s) \, f_2 \begin{pmatrix} s, u_1(s), \ldots, u_1^{(m_2)}(s), \\ u_2(s), \ldots, u_2^{(n_2)}(s) \end{pmatrix} ds.
\end{cases}
\tag{11.3}
$$

Next, we will show that the operator $T : E \to E$ defined by $T = (T_1, T_2)$ has a fixed point on E.

The proof will follow several steps, presented in detail for operator $T_1(u, v)$. The technique for operator $T_2(u, v)$ is similar.

Step 1: *T is well defined and uniformly bounded in E.*

Consider a bounded set $D \subseteq E$ and $(u, v) \in D$. Therefore, there is $\rho_1 > 0$ such that

$$\|(u, v)\|_E \leq \rho_1. \tag{11.4}$$

By the Lebesgue Dominated Theorem, $(A1)$, $(A2)$ and because f_1 is a L^∞-Carathéodory function, follow that, for $i = 0, 1, \ldots, r_1$,

$$
\begin{aligned}
\|T_1(u_1, u_2)^{(i)}\|_\infty &= \sup_{t \in \mathbb{R}} \left| \int_{-\infty}^{+\infty} \frac{\partial^i k_1}{\partial t^i}(t, s) \, g_1(s) \, f_1 \begin{pmatrix} s, u_1(s), \ldots, u_1^{(m_1)}(s), \\ u_2(s), \ldots, u_2^{(n_1)}(s) \end{pmatrix} ds \right| \\
&\leq \int_{-\infty}^{+\infty} \phi_{1i}(s) g_1(s) \left| f_1 \begin{pmatrix} s, u_1(s), \ldots, u_1^{(m_1)}(s), \\ u_2(s), \ldots, u_2^{(n_1)}(s) \end{pmatrix} \right| ds \\
&\leq \int_{-\infty}^{+\infty} \phi_{1i}(s) \varphi_{1\rho_1}(s) g_1(s) \, ds < +\infty.
\end{aligned}
$$

Taking into account these arguments, T_2 verifies similar bounds and $\|T(u, v)\|_E < +\infty$, that is $TD \subseteq E$ is uniformly bounded.

Step 2: *T is equicontinuous in E.*

Consider $t_1, t_2 \in \mathbb{R}$ and suppose without loss of generality, that $t_1 \leq t_2$. So, by $(A1)$, for $i = 0, 1, \ldots, r_1$,

$$
\begin{aligned}
&|(T_1(u_1, u_2))^{(i)}(t_1) - (T_1(u_1, u_2))^{(i)}(t_2)| \\
&\leq \int_{-\infty}^{+\infty} \left| \frac{\partial^i k_1}{\partial t^i}(t_1, s) - \frac{\partial^i k_1}{\partial t^i}(t_2, s) \right| |g_1(s)| \left| f_1 \begin{pmatrix} s, u_1(s), \ldots, u_1^{(m_1)}(s), \\ u_2(s), \ldots, u_2^{(n_1)}(s) \end{pmatrix} \right| ds \\
&\leq \int_{-\infty}^{+\infty} \left| \frac{\partial^i k_1}{\partial t^i}(t_1, s) - \frac{\partial^i k_1}{\partial t^i}(t_2, s) \right| g_1(s) \, \varphi_{1\rho_1}(s) ds \to 0, \quad \text{as } t_1 \to t_2.
\end{aligned}
$$

Therefore, $T_1 D$ is equicontinuous in E_1. In the same way it can be shown that $T_2 D$ is equicontinuous on E_2. Thus, TD is equicontinuous on E.

Step 3: *TD is equiconvergent at $\pm\infty$.*

Consider $(u, v) \in D \subseteq E$ and $i = 0, 1, \ldots, r_1$. For the operator T_1, we have by $(A1)$,

$$
\begin{aligned}
&|(T_1(u_1, u_2))^{(i)}(t) - \lim_{t \to \pm\infty} (T_1(u_1, u_2))^{(i)}(t)| \\
&\leq \int_{-\infty}^{+\infty} \left| \frac{\partial^i k_1}{\partial t^i}(t, s) - \frac{\partial^i k_1}{\partial t^i}(\pm\infty, s) \right| |g_1(s)| \left| f_1 \begin{pmatrix} s, u_1(s), \ldots, u_1^{(m_1)}(s), \\ u_2(s), \ldots, u_2^{(n_1)}(s) \end{pmatrix} \right| ds \\
&\leq \int_{-\infty}^{+\infty} \left| \frac{\partial^i k_1}{\partial t^i}(t, s) - \frac{\partial^i k_1}{\partial t^i}(\pm\infty, s) \right| g_1(s) \, \varphi_{1\rho_1}(s) ds \to 0, \quad \text{as } t \to \pm\infty.
\end{aligned}
$$

$T_1 D$ is equiconvergent at $\pm\infty$. By similar arguments, it can be proved that $T_2 D$ is equiconvergent at $\pm\infty$. Moreover, TD is equiconvergent at $\pm\infty$. By Lemma 3.1, TD is relatively compact.

Step 4: *$T\Omega \subset \Omega$ for some $\Omega \subset E$ a closed, bounded and convex set.*

Consider

$$\Omega := \{(u,v) \in E : \|(u,v)\|_E \leq \rho_2\},$$

with $\rho_2 > 0$ such that, for $\iota = 1, 2$ and $i = 0, \ldots, r_\iota$,

$$\rho_2 := \max\left\{\rho_1, \int_{-\infty}^{+\infty} \phi_{\iota j}(s)g_\iota(s)\varphi_{\rho_1}(s)ds\right\}$$

with $\rho_1 > 0$ given by (11.4).

Following the arguments used in Step 1 we have, for $(u,v) \in \Omega$,

$$\|T(u,v)\|_E = \|(T_1(u,v), T_2(u,v))\|_E$$
$$= \max\left\{\|T_1(u,v)\|_{E_1}, \|T_2(u,v)\|_{E_2}\right\} \leq \rho_2.$$

So, $T\Omega \subset \Omega$, and by Theorem 5.2, the operator $T(u,v) = (T_1(u,v), T_2(u,v))$, has a fixed point $(u,v) \in E_1 \times E_2$, that is, the problem (11.1) has at least one solution. □

Remark 11.1. If, for $\iota = 1, 2$,

$$\lim_{t \to -\infty} k_\iota(t,s) = \lim_{t \to +\infty} k_\iota(t,s),$$

then the solution of (11.1) is a homoclinic solution, otherwise it is a heteroclinic solution.

11.3 Application to fourth-order coupled systems of infinite beams deflection model

Recently, in [125], the authors study the two-beam coupled structure as two infinite beams and considering the coupling between the bending wave and the torsion, the conversion of wave types at the coupled interface, as well as other details on the coupling of beams.

Motivated by the concept of very large floating structures and ice plates in waves, in [135], Jang *et al.* consider the inverse loading distribution from measured deflection of an infinite beam on elastic foundation. They express the relationship between the loading distribution and vertical deflection of the beam in the form of an integral equation of the first kind.

An efficient method for the static deflection analysis of an infinite beam on a nonlinear elastic foundation is developed in [11], where the authors combine the quasilinear method and Green's functions to obtain the approximate solution.

Motivated by the works above and more concretely in [134], where the authors analyze moderately large deflections of infinite nonlinear beams resting on elastic foundations under localized external loads, we consider an arbitrary family of nonlinear coupled systems of Bernoulli–Euler–v. Karman problem composed by the fourth-order coupled system

$$
\begin{cases}
E_1 I_1 u^{(4)}(x) + \eta_1 u(x) = \frac{1}{(1+x^2)^2} \left[\frac{3}{2} E_1 A_1 \left(u'(x) \right)^2 v''(x) + \omega_1(x) \right], \ x \in \mathbb{R} \\[2ex]
E_2 I_2 v^{(4)}(x) + \eta_2 v(x) = \frac{1}{(1+x^2)^2} \left[\frac{3}{2} E_2 A_2 \left(v'(x)u'(x) \right)^2 u''(x) + \omega_2(x) \right]
\end{cases}
$$

$$(11.5)$$

and the boundary conditions

$$
\begin{cases}
u(\pm\infty) = 0, \ v(\pm\infty) = 0, \\
u'(\pm\infty) = 0, \ v'(\pm\infty) = 0,
\end{cases}
$$

$$(11.6)$$

where,

$$
w(\pm\infty) := \lim_{x \to \pm\infty} w(x).
$$

We also emphasize that:

- E_i, I_i, $i = 1, 2$ are the Young's modulus (the elastic modulus of the material) and the mass moment of inertia of the cross section of beam 1 and beam 2, respectively;
- $\eta_1 u(x)$, $\eta_2 v(x)$ are the spring forces upward of the first and second beams, respectively;
- A_1, A_2 are the cross-sectional areas of the first and second beams, respectively;
- $\omega_1(x)$, $\omega_2(x)$ are the positive localized applied loading downward on the corresponding beams.

In fact, the differential system (11.5)-(11.6) can be rewritten as the following system of integral equations

$$
\begin{cases}
u(x) = \int_{-\infty}^{+\infty} k_1(x, s) \, \frac{1}{E_1 I_1} \frac{1}{(1 + s^2)^2} \\
\qquad \times \left[\frac{3}{2} E_1 A_1 \left(u'(s) \right)^2 v''(s) + \omega_1(s) + \eta_1 u(s) \right] ds, \\[2ex]
v(x) = \int_{-\infty}^{+\infty} k_2(x, s) \, \frac{1}{E_2 I_2} \frac{1}{(1 + s^2)^2} \\
\qquad \times \left[\frac{3}{2} E_2 A_2 \left(v'(s)u'(s) \right)^2 u''(s) + \omega_2(s) + \eta_2 v(s) \right] ds,
\end{cases}
$$

$$(11.7)$$

where the kernel functions $k_1(x,s)$ and $k_2(x,s)$ are given, respectively, by the corresponding Green's functions

$$k_\iota(x,s) = \frac{e^{-\alpha_\iota|s-x|}}{\sqrt{2}^5 \alpha_\iota^3} \sin\left(\alpha_\iota|s-x| + \frac{\pi}{4}\right), \tag{11.8}$$

with $\alpha_\iota = \frac{\sqrt{2}}{2} \sqrt[4]{\frac{\eta_\iota}{E_\iota I_\iota}}$ *for* $\iota = 1,2$.

For $j = 0,1,2$, $\iota = 1,2$, and defining

$$k_{\iota j}^-(x,s) := \frac{e^{\alpha_\iota(s-x)}}{\sqrt{2}^{5-j} \alpha_\iota^{3-j}} \sin\left(\alpha_\iota(x-s) + \frac{\pi(3j+1)}{4}\right),$$

$$k_{\iota j}^+(x,s) := \frac{e^{\alpha_\iota(x-s)}}{\sqrt{2}^{5-j} \alpha_\iota^{3-j}} \sin\left(\alpha_\iota(s-x) + \frac{\pi(3j+1)}{4}\right),$$

we have

$$u^{(j)}(x) = \int_{-\infty}^x k_{1j}^-(x,s)F_1(s)ds + (-1)^j \int_x^{+\infty} k_{1j}^+(x,s)F_1(s)ds, \tag{11.9}$$

and

$$v^{(j)}(x) = \int_{-\infty}^x k_{2j}^-(x,s)F_2(s)ds + (-1)^j \int_x^{+\infty} k_{2j}^+(x,s)F_2(s)ds,$$

with

$$F_1(s) = \frac{1}{1+s^2}\left[\frac{3}{2}E_1 A_1\left(u'(s)\right)^2 v''(s) + \omega_1(s) + \eta_1 u(s)\right],$$

$$F_2(s) = \frac{1}{1+s^2}\left[\frac{3}{2}E_2 A_2\left(v'(s)u'(s)\right)^2 u''(s) + \omega_2(s) + \eta_2 v(s)\right].$$

The system (11.7) is a particular case of (11.1) with $r_1 = r_2 = m_1 = m_2 = n_1 = n_2 = 2$, $g_1(x) = \frac{1}{1+x^2}\frac{1}{E_1 I_1}$, $g_2(x) = \frac{1}{1+x^2}\frac{1}{E_2 I_2}$, $E_1 I_1 > 0$, $E_2 I_2 > 0$ and

$$f_1(x,y_1,\ldots,y_6) = \frac{1}{1+x^2}\left[\frac{3}{2}E_1 A_1 y_2^2 y_6 + \omega_1(x) + \eta_1 y_1\right],$$

$$f_2(x,y_1,\ldots,y_6) = \frac{1}{1+x^2}\left[\frac{3}{2}E_2 A_2\left(y_5 y_2\right)^2 y_3 + \omega_2(x) + \eta_2 y_4\right].$$

The functions f_1, $f_2 : \mathbb{R}^7 \to \mathbb{R}$, respectively, are L^∞-Carathéodory functions, as, for $\rho > 0$ such that

$$\max\{|y_1|, |y_2|, |y_3|, |y_4|, |y_5|, |y_6|\} < \rho,$$

there exist functions $\varphi_{1\rho}, \varphi_{2\rho} \in L^{\infty}(\mathbb{R})$ verifying

$$f_1(x, y_1, \ldots, y_6) \leq \frac{1}{1+x^2}\left[\frac{3}{2}E_1 A_1 \rho^3 + \sup_{x \in \mathbb{R}} \omega_1(x) + \eta_1 \rho\right] := \varphi_{1\rho}(x),$$

$$f_2(x, y_1, \ldots, y_6) \leq \frac{1}{1+x^2}\left[\frac{3}{2}E_2 A_2 \rho^3 + \sup_{x \in \mathbb{R}} \omega_2(x) + \eta_2 \rho\right] := \varphi_{2\rho}(x).$$

Note also that (A1) and (A2) are satisfied, since, for $\iota = 1, 2$ and $j = 0, 1, 2$,

$$\lim_{x \to \pm\infty} \frac{\partial^j k_\iota}{\partial x^j}(x, s) = 0,$$

$$\left|\frac{\partial^j k_\iota}{\partial x^j}(x, s)\right| \leq \frac{1}{\sqrt{2}^{5-j} \alpha_\iota^{3-j}} := \phi_{\iota j}, \ \forall s \in \mathbb{R},$$

and

$$\int_{-\infty}^{+\infty} \phi_{\iota j}(s) g_\iota(s) \varphi_{\iota\rho}(s) ds < +\infty.$$

So, by Theorem 11.1, there is at least a nontrivial solution $(u, v) \in E_1 \times E_2$ of problem (11.5)-(11.6). In addition, by Remark 11.1 the solution is a nontrivial homoclinic solution.

Conclusions and open issues

The last Part has two main goals: The first one, is to consider coupled systems with integral equations as one way to study higher-order boundary value problems based on differential equations, but requiring less regularity on the nonlinear parts. The second one is to present some techniques for the solvability of Hammerstein-integral equations where the kernel functions have fixed sign or change of sign, on bounded or unbounded intervals.

Fixed point theory seems to be an adequate method to deal with the integral coupled systems where the nonlinearities depend on several derivatives of both variables and, moreover, the derivatives can be of different order on each variable and each equation. In this work we apply the Guo-Krasnosel'skiĭ fixed-point theorem for expansive and compressive cones, to obtain positive solutions, and Schauder's fixed point theorem for operators with negative kernels, or where the kernel functions can, locally, change sign.

The above research has led to some open questions:

- In Chapters 9 and 10, can the asymptotic growth assumption pairs (B1) and (B2) be replaced for a unique condition near the origin or at $+\infty$?

- Is there any relation between the solutions of the coupled integral system (11.1) on the real line, and homoclinic and/or heteroclinic solutions?

- Are there other advantages, besides the regularity, to transform boundary value problems in their integral form?

Bibliography

[1] N.H. Abel, *Solutions de quelques problèmes à l'aide d'intégrales défines*, Oeuvres complètes, nouvelle éd., 1, Grondahl & Son, Christiania (1881) pp. 11–27 (Edition de Holmboe).

[2] A. Abdel-Rady, A. El-Sayed, S. Rida, I. Ameen, *On some impulsive differential equations*, Math. Sci. Lett. 1 No. 2, (2012) 105–113.

[3] E. Abdolmaleki, H Najafi, *An efficient algorithmic method to solve Hammerstein integral equations and application to a functional differential equation*, Advances in Mechanical Engineering Vol. 9(6) (2017) 1–8, DOI: 10.1177/1687814017701704.

[4] R. Agarwal, D. O'Regan, P. Wong, *Constant-Sign Solutions of Systems of Integral Equations*, Springer International Publishing Switzerland, 2013.

[5] R. Agarwal, J. Banaś, K. Banaś, D. O'Regan, *Solvability of a quadratic Hammerstein integral equation in the class of functions having limits at infinity*, Journal of Integral Equations and Applications, Vol. 2, No. 2, 2011.

[6] R. Agarwal, D. O'Regan, *Infinite Interval Problems for Differential, Difference and Integral Equations*, Kluwer Academic Publisher, Glasgow, 2001.

[7] R. Agarwal, D. O'Regan, *Integral and Integrodifferential Equations: Theory, Methods and Applications*, Gordon and Breach Science Publishers, Amsterdam, 2000.

[8] M. Aguiar, P. Ashwin, A. Dias, M. Field, *Dynamics of coupled cell networks: synchrony, heteroclinic cycles and inflation*, Journal of Nonlinear Science 21 (2), (2011) pp. 271–323.

[9] B. Ahmad, J. Nieto, *Existence and approximation of solutions for a class of nonlinear impulsive functional differential equations with anti-periodic boundary conditions*, Nonlinear Analysis 69 (2008) 3291–3298.

[10] B. Ahmad, *Existence of solutions for second-order nonlinear impulsive boundary-value problems*, Electronic Journal of Differential Equations, Vol. 2009 (2009), No. 68, pp. 1–7.

[11] F. Ahmada, M. Ullahb, T. Jangc, E. Alaidarousb, *An efficient method for the static deflection analysis of an infinite beam on a nonlinear elastic foundation of one-way spring model*, Ships and Offshore Structures, (2014), http://dx.doi.org/10.1080/17445302.2014.956381.

[12] A. Algaba, E. Freire, E. Gamero, A. Rodríguez-Luis, *An exact homoclinic orbit and its connection with the Rössler system*, Physics Letters A 379 (2015) 1114–1121.

[13] M. Almousa, A. Ismail, *Numerical solution of Fredholm-Hammerstein integral equations by using optimal homotopy asymptotic method and homotopy perturbation method*, AIP Conference Proceedings 1605, 90 (2014); doi: 10.1063/1.4887570.

[14] C. D. Aliprantis, K. C. Border, *Infinite Dimensional Analysis*, 3rd ed., Springer-Verlag, Berlin (2005).

[15] C. Alves, *Existence of heteroclinic solution for a class of non-autonomous second-order equation*, Nonlinear Differ. Equ. Appl. 22 (5) (2015) 1195–1212.

[16] H. Amann, *Parabolic evolution and nonlinear boundary conditions*, J. Differential Equations, 72 (1988) 201–269.

[17] N. Ambraseys, *The seismic stability of earth dams*, 2nd WCEE, Tokyo, pp. 1345–1363, 1960.

[18] Y. An, *Nonlinear perturbations of a coupled system of steady state suspension bridge equations*, Nonlinear Analysis **51** (2002) 1285–1292.

[19] J. Andres, G. Gabor, L. Gorniewicz, *Boundary Value Problems on Infinite Intervals*, Transactions of the American Mathematical Society, 351 (12) (1999) 4861–4903, S 0002-9947(99)02297-7.

[20] S. Ariaratnam, D. Tam and W-C. Xie, *Lyapunov Exponents of Two-Degrees-of-Freedom Linear Stochastic Systems*, Stochastic Structural Dynamics 1, New Theoretical Developments, Springer-Verlag, Berlin, 1991.

[21] R. Aris, *Introduction to the analysis of chemical reactors*, Prentice-Hall, Englewood Cliffs, NJ, (1965).

[22] D.G. Aronson, *A comparison method for stability analysis of nonlinear parabolic problems*, SIAM Rev., 20, (1978).

[23] P. Ashwin, Ö. Karabacak, *Robust Heteroclinic Behaviour, Synchronization, and Ratcheting of Coupled Oscillators*, Dynamics, Games and Science II, (2011) pp. 125–140.

[24] P. Ashwin, G. Orosz, J. Wordsworth, S. Townley, *Dynamics on Networks of Cluster States for Globally Coupled Phase Oscillators*, SIAM J. Appl. Dynam. Syst. Vol. 6, No. 4, pp. 728–758.

[25] N.A. Asif, R.A. Khan, *Positive solutions to singular system with four-point coupled boundary conditions*, J. Math. Anal. Appl. **386** (2012) 848–861.

[26] N.A. Asif, P.W. Eloe, R. A. Khan, *Positive solutions for a system of singular second order nonlocal boundary value problems*, J. Korean Math. Soc., 47, no. 5 (2010) 985–1000.

[27] K. Atkinson, *The Numerical Solution of Integral Equations of the Second Kind*, Cambridge Monographs on Applied and Computational Mechanics, New York, 1997.

[28] C. Bai, J. Fang, *On positive solutions of boundary value problems for second-order functional differential equations on infinite intervals*, J. Math. Anal. Appl. 282 (2003) 711–731.

[29] D. Bainov, A. Dishliev, *Population dynamics control in regard to minimizing the time necessary for the regeneration of a biomass taken away from the population*, Comp. Rend. Bulg. Sci., 42 (1989) 29–32.

[30] D. Bainov, V. Lakshmikantham, P. Simeonov, *Theory of Impulsive Differential Equations*, Series in Modern Applied Mathematics, World Scientific, Singapore, 6 (1989).

[31] D. Bainov, P. Simeonov, *Impulsive differential equations: periodic solutions and applications*, Longman Scientific and Technical, Essex, England, (1993).

[32] D. Bainov, P. Simeonov, *Oscillation Theory of Impulsive Differential Equations*, International Publications, Orlando, Fla, USA (1998).

[33] H. Bakodah, M. Darwish, *Solving Hammerstein Type Integral Equation by New Discrete Adomian Decomposition Methods*, Hindawi Publishing Corporation Mathematical Problems in Engineering Volume 2013, Article ID 760515, 5 pages http://dx.doi.org/10.1155/2013/760515.

[34] V. Balakrishnan, *All about the Dirac Delta Function(?)*, Resonance Vol. 8-8, (2003) 48–58.

[35] J. Banaś, K. Sadarangani, *Compactness condition in the study of functional, differential and integral equations*, Rev. Mat. Univ. Complut. Madrid, v. 2, n. 1, pp. 31–38, 2013.

[36] J. Banaś, J. Martin, K. Sadarangani, *On solutions of a quadratic integral equation of Hammerstein type*, Mathematical and Computer Modelling 43 (2006) 97–104.

[37] A. Barzkar, P. Assari, M. Mehrpouya, *Application of the CAS Wavelet in Solving Fredholm-Hammerstein Integral Equations of the Second Kind with Error Analysis*, World Applied Sciences Journal (2012) (18) (12): 1695–1704, DOI: 10.5829/idosi.wasj.2012.18.12.467.

[38] A. Bassett, A. Krause, R. Gorder, *Continuous dispersal in a model of predator-prey-subsidy population dynamics*, Ecological Modelling 354 (2017) 115–122.

[39] J. Baxley, *Existence and uniqueness for nonlinear boundary value problems on infinite intervals*, Journal of Mathematical Analysis and Applications, vol. 147, (1990) 122–133.

[40] J. Beitia, V. Garcia, P. Torres, *Solitary Waves for Linearly Coupled Nonlinear Schrödinger Equations with Inhomogeneous Coefficients*, J. Nonlinear Sci. (2009) 19: 437. doi:10.1007/s00332-008-9037-7.

[41] M. Benchohra, B. Slimani, *Existence and uniqueness of solutions to impulsive fractional differential equations*, Electron. J. Differ. Equ., (10), (2009) 1–11.

[42] M. Benchohra, J. Henderson, S. Ntouyas, *Impulsive Differential equations and Inclusions*, Hindawi Publishing Corporation, New York (2006).

[43] F. Bernis, L.A. Peletier, *Two problems from draining flows involving third-order ordinary differential equations*, SIAM J. Math. Anal. **27** (2) (1996) 515–527.

[44] A. Bertozzi, *Heteroclinic orbits and chaotic dynamics in planar fluid flows*, Siam J. Math. Anal. Vol. 19, No. 6, November 1988.

[45] A. Bica, M. Curila, S. Curila, *About a numerical method of successive inter-polations for functional Hammerstein integral equations*, Journal of Computational and Applied Mathematics 236 (2012) 2005–2024.

[46] W. Boyce, R. DiPrima, *Elementary Differential Equations and Boundary Value Problems*, 7th ed., John Wiley & Sons, Inc., New York, 2001.

[47] H. Brunner, Z. W. Yang, *Blow-up behavior of Hammerstein-type Volterra integral equations* , Journal of Integral Equations and Applications Volume 24, Number 4, Winter 2012.

[48] A. Buryak, Y. Kivshar, *Twin-hole dark solitons*, Phys. Rev. A 51 (1995) R41–R44.

[49] A. Buryak, Y. Kivshar, *Solitons due to second harmonic generation*, Physics Letters A 197 (1995) 407–412.

[50] A. Cabada, F. Minhós, *Fully nonlinear fourth-order equations with functional boundary conditions*, J. Math. Anal. Appl. 340 (2008) 239–251.

[51] A. Cabada, M. R. Grossinho, F. Minhós, *On the Solvability of some Discontinuous Third Order Nonlinear Differential Equations with Two Point Boundary Conditions*, J. Math. Anal. Appl., 285 (2003) 174–190.

[52] A. Cabada, J. Cid, *Heteroclinic solutions for non-autonomous boundary value problems with singular Φ-Laplacian operators*, Discret and Continuous Dynamical Systems Supplement 2009, 118–122.

[53] A. Cabada, J. Cid, G. Infante, *A positive fixed point theorem with applications to systems of Hammerstein integral equations*, Boundary Value Problems 2014:254, https://doi.org/10.1186/s13661-014-0254-8.

[54] A. Cabada, J. Fialho, F. Minhós, *Non ordered lower and upper solutions to fourth order functional BVP*, Discrete Contin. Dyn. Syst., Supplements Vol. 2011, Special Issue (2011) 209–218.

[55] A. Cabada, L. López-Somoza, F. Tojo, *Existence of solutions of integral equations defined in unbounded domains via spectral theory*, https://arxiv.org/abs/1811.06121 (2018).

[56] A. Cabada, L. López-Somoza, F. Tojo, *Existence of solutions of integral equations with asymptotic conditions*, Nonlinear Analysis: Real World Applications 4 (2018) 140–159.

[57] A. Calamai, *Heteroclinic solutions of boundary value problems on the real line involving singular Φ-Laplacian operators*, J. Math. Anal. Appl. 378 (2011) 667–679.

[58] T. Cardinali, N. Papageorgiou, *Hammerstein integral inclusions in reflexive Banach spaces*, Proceedings of the American Mathematical Society, Volume 127, Number 1, January 1999, pp. 95–103.

[59] A. Catlla, D. Schaeffer, T. Witelski, E. Monson, A. Lin, *On Spiking Models for Synaptic Activity and Impulsive Differential Equations*, Society for Industrial and Applied Mathematics, Vol. 50, No. 3, (2008) 553–569.

[60] S. Chandler-Wilde, M. Lindner, *Boundary integral equations on unbounded rough surfaces: Fredholmness and the finite section method*, The Journal of Integral Equations and Applications Vol. 20, No. 1 (Spring 2008), pp. 13–48.

[61] A. Champneys, G. Lord, *Computation of homoclinic solutions to periodic orbits in a reduced water-wave problem*, Physica D 102 (1997) 101–124.

[62] A. Champneys, *Homoclinic orbits in reversible systems and their applications in mechanics, fluids and optics*, Physica D 112 (1998) 158–186.

[63] P. Chen, X. Tang, *New existence of homoclinic orbits for a second-order Hamiltonian system*, Computers and Mathematics with Applications 62 (2011) 131–141.

[64] A. Cheng, D. Cheng, *Heritage and early history of the boundary element method*, Engineering Analysis with Boundary Elements 29 (2005) 268–302.

[65] K. Church, *Applications of Impulsive Differential Equations to the Control of Malaria Outbreaks and Introduction to Impulse Extension Equations: A General Framework to Study the Validity of Ordinary Differential Equation Models with Discontinuities in State*, https://books.google.pt/books?id=7bipAQAACAAJ, 2014.

[66] F. Cianciaruso, G. Infante, P. Pietramala, *Solutions of perturbed Hammerstein integral equations with applications*, Nonlinear Analysis: Real World Applications 33 (2017) 317–347.

[67] M. Coclite, *Positive solutions of Hammerstein integral equation with singular nonlinear term*, Topological Methods in Nonlinear Analysis Journal of the Juliusz Schauder Center Vol. 15 (2000) 235–250.

[68] G. Coclite, M. Coclite, *Elliptic Perturbations for Hammerstein Equations with Singular Nonlinear Term*, Electron. J. Diff. Eqns. 2006 (2006) no. 104, 1–23.

[69] G. Coclite, M. Coclite, *Positive solutions for an integro-differential equation with singular nonlinear term*, Differential Integral Equations 18 (2005) no. 9, 1055–1080.

[70] G. Coclite, M. Coclite, *Stationary solutions for conservation laws with singular nonlocal sources*, J. Differential Equations 248 (2010) no. 2, 229–251.

[71] G. M. Coclite, M. M. Coclite, *Conservation laws with singular nonlocal sources*, J. Differential Equations 250 (2011) no. 10, 3831–3858.

[72] Y. Cui, J. Sun, *On existence of positive solutions of coupled integral boundary value problems for a nonlinear singular superlinear differential system*, Electron. J. Qual. Theory Differ. Equ. 41 (2012) 1–13.

[73] H. Curtis, *Orbital Mechanics for Engineering Students*, Biddles Ltd, King's Lynn, Norfolk, 2005.

[74] L. Danziger, G. Elmergreen, *The thyroid-pituitary homeostatic mechanism*, Bull. Math. Biophys. **18** (1956) 1–13.

[75] A. Dishliev, D. Bainov, *Dependence upon initial conditions and parameters of solutions of impulsive differential equations with variable structure*, International Journal of Theoretical Physics 29 (1990), 655–676.

[76] A. Dishliev, K. Dishlieva, S. Nenov, *Specific Asymptotic Properties of the Solutions of Impulsive Differential Equations. Methods and Applications*, Academic Publications, Ltd., 2012.

[77] K. Dishlieva, *Impulsive Differential Equations and Applications*, J Applied Computat Mathemat 1(6) (2012), http://dx.doi.org/10.4172/2168-9679.1000e117.

[78] S. Djebali, Z. Sahnoun, *Nonlinear alternatives of Schauder and Krasnosel'skij types with applications to Hammerstein integral equations in L^1 spaces,* Journal of Differential Equations, 249(9) (2010), 2061-2075.

[79] A. D'onofrio, *On pulse vaccination strategy in the SIR epidemic model with vertical transmission,* Appl. Math. Lett., 18 (2005) 729–732.

[80] E. Ellero, F. Zanolin, *Homoclinic and heteroclinic solutions for a class of second-order non-autonomous ordinary differential equations: multiplicity results for stepwise potentials,* Boundary Value Problems 2013 (2013) 167. https://doi.org/10.1186/1687-2770-2013-167.

[81] P. Eloe, E. Kaufmann, C. Tisdell, *Multiple solutions of a boundary value problem on an unbounded domain,* Dyn. Syst. Appl. 15(1), (2006) 53–63.

[82] P. Eloe, E. Kaufmann, C. Tisdell, *Existence of solutions for second-order differential equations and systems on infinite intervals,* Electronic Journal of Differential Equations, Vol. 2009 (2009), No. 94, pp. 1–13.

[83] P. Eloe, J. Henderson, B. Thompson, *Extremal Points for Impulsive Lidstone Boundary Value Problems,* Mathematical and Computer Modelling 32 (2000) 687–698.

[84] E. L. Ence, *Ordinary Differential Equations,* 1st edn., Dover Publications, Inc., New York (1956).

[85] L. Erbe, X. Liu, *Existence results for boundary value problems of second order impulsive differential equations,* J. Math. Anal. Appl., 149 (1990) 56–69.

[86] G. Ermentrout, D. Terman, *Mathematical Foundations of Neuroscience,* Springer-Verlag, New York, 2010.

[87] F. Faraci, V. Moroz, *Solutions of Hammerstein Integral Equations Via Variational Principle,* Journal of Integral Equations and Applications Volume 15, Number 4, Winter 2003.

[88] R. Farengo, Y. Lee, P. Guzdar, *An electromagnetic integral equation: Application to microtearing modes,* Physics of Fluids (1983) (26) (12), doi:10.1063/1.864112.

[89] T. H. Fay, S. D. Graham, *Coupled spring equations,* Int. J. Math. Educ. Sci. Technol., Vol. 34, No. 1 (2003) 65–79.

[90] M. Fečkan, *Bifurcation and Chaos in Discontinuous and Continuous Systems,* Springer-Verlag, Berlin, 2011.

[91] B. Feng, R. Hu, *A Survey On Homoclinic And Heteroclinic Orbits,* Applied Mathematics E-Notes, 3(2003), 16–37.

[92] J. Fialho, F. Minhós, *Higher order periodic impulsive problems,* Dynamical Systems, Differential Equations and Applications, AIMS Proceedings, (2015) 446-454.

[93] J. Fialho, F. Minhós, *Fourth Order Impulsive Periodic Boundary Value Problems,* Dynamical Systems, Differential Equations and Applications, AIMS Proceedings (2013).

[94] P. Fife, *Boundary and Interior Transition Layer Phenomena for Pairs of Second-Order Differential Equations,* J. Math. Anal. Appl. 54 (1976) 497–521.

[95] R. Figueroa, F. Tojo, *Fixed points of Hammerstein-type equations on general cones*, Fixed Point Theory 19(2) (2016), DOI: 10.24193/fpt-ro.2018.2.45.

[96] M. Frigon, D. O'Regan, *Impulsive differential equations with variable times*, Nonlinear Anal. 26 (1996) 1913–1922.

[97] F. Gazzola, *Mathematical Models for Suspension Bridges: Nonlinear Structural Instability*, Modeling, Simulations and Applications, Vol. 15, Springer, 2015.

[98] J. Graef, L. Kong, F. Minhós, J. Fialho, *On the lower and upper solution method for higher order functional boundary value problems*, Appl. Anal. Discrete Math. 5,(1) (2011) 133–146.

[99] J. Graef, L. Kong, F. Minhós, *Higher order boundary value problems with Φ-Laplacian and functional boundary conditions*, Comput. Math. Appl. 61 (2011) 236–249.

[100] J. Graef, L. Kong, F. Minhós, *Generalized Hammerstein equations and applications*, Results Math. 72 (2017) 369–383.

[101] S. Grigolato, S. Panizza, M. Pellegrini, P. Ackerman, R. Cavalli, *Light-lift helicopter logging operations in the Italian Alps: a preliminary study based on GNSS and a video camera system*, Forest Science and Technology 12(2) 88–97, DOI: 10.1080/21580103.2015.1075436.

[102] M. R. Grossinho, F. Minhós, A. I. Santos, *A note on a class of problems for a higher order fully nonlinear equation under one sided Nagumo type condition*, Nonlinear Anal. **70** (2009) 4027–4038.

[103] A. Guezane–Lakoud, L. Zenkoufi, *Study of Third-Order Three-Point Boundary Value Problem with Dependence on the First-Order Derivative*, Advances in Applied Mathematics and Approximations Theory, Springer Proceedings in Mathematics & Statistics 41, DOI:10.1007/978-1-4614-6393-128, Springer Science+Business Media New York 2013.

[104] D. Guo, *Boundary value problems for impulsive integro-differential equation on unbounded domains in a Banach Space*, Appl. Math. Comput. 99 (1999) 1–15.

[105] D. Guo, *A class of second-order impulsive integro-differential equations on unbounded domain in Banach space*, Applied Mathematics and Computations 125 (2002) 59–77.

[106] D. Guo, V. Lakshmikantham, X. Liu, *Nonlinear Integral Equations in Abstract Spaces*, Kluwer Academic Publishers, Boston, MA, 1996.

[107] D. Guo, V. Lakshmikantham, *Nonlinear problems in abstract cones,*. Academic Press, 1988.

[108] G. Guseinov, I. Yaslan, *Boundary value problems for second order nonlinear differential equations on infinite intervals*, J. Math. Anal. Appl. 290 (2004) 620–638.

[109] G. Gustafson, *Applied Differential Equations and Linear Algebra*, to appear (https://faculty.utah.edu/u0031553-GRANT_B_GUSTAFSON/bibliography/index.html).

[110] A. Hammerstein, *Nichtlineare intergralgleichungen nebst anwendungen*, Acta Math. 54 (1929), 117–176.

[111] Z. Hao, J. Liang, T. Xiao, *Positive solutions of operator equations on half-line*, Journal of Mathematical Analysis and Applications, vol. 314, (2006) 423–435.

[112] S. Hasselmann, K. Hasselmann, G.K. Komen, P. Janssen, A.J. Ewing, V. Cardone, *The WAM model-A third generation ocean wave prediction model* Journal of Physical Oceanography, 18, pp. 1775–1810, 1988.

[113] J. Henderson, R. Luca, *Positive solutions for a system of higher-order multi-point boundary value problems*, Comput. Math. Appl. 62 (2011) 3920–3932.

[114] J. Henderson, R. Luca, *Positive solutions for a system of second-order multi-point boundary value problems*, Appl. Math. Comput. 218 (2012) 6083–6094.

[115] J. Henderson, S.K. Ntouyas, *Positive solutions for systems of nth order three-point nonlocal boundary value problems*, Electron. J. Qual. Theor. Differ.Equat. (18) (2007) 1–12.

[116] J. Henderson, S.K. Ntouyas, *Positive solutions for systems of nonlinear boundary value problems*, Nonlinear Stud. 15 (2008) 51–60.

[117] J. Henderson, S.K. Ntouyas, *Positive solutions for systems of three-point nonlinear boundary value problems*, Aust. J. Math. Anal. Appl. 5 (2008) 1–9.

[118] J. Henderson, S.K. Ntouyas, I. Purnaras, *Positive solutions for systems of generalized three-point nonlinear boundary value problems*, Comment. Math.Univ. Carolin. 49 (2008) 79–91.

[119] J. Henderson, S.K. Ntouyas, I.K. Purnaras, *Positive solutions for systems of m-point nonlinear boundary value problems*, Math. Model. Anal. 13 (2008) 357–370.

[120] J. Henderson, R. Luca, *Boundary Value Problems for Systems of Differential, Difference and Fractional Equations, Positive Solutions*, Elsevier, 2015.

[121] J. Henderson, R. Luca, *Positive solutions for systems of nonlinear second-order multipoint boundary value problems*, Math. Methods Appl. Sci. , **37** (2014) 2502–2516.

[122] A. Homburg, B. Sandstede, *Homoclinic and Heteroclinic Bifurcations in Vector Fields*, Handbook of Dynamical Systems Vol. 3 (2010), pp. 379-524, doi: 10.1016/S1874-575X(10)00316-4.

[123] T. Horikis, *Modulation instability and solitons in two-color nematic crystals*, Physics Letters A 380 (2016) 3473–3479.

[124] C. Hsu, W. Cheng, *Applications of the theory of impulsive parametric excitation and new treatments of general parametric excitation problems*, Trans. ASME J. Appl. Mech., 40 (1973) 551–558.

[125] W. Hu, H. Chen, Q. Zhong, *Research on transmission coefficient property of coupled beam structure*, Vibroengineering Procedia 20 (2018).

[126] L. Hu, L. Liu, Y. Wu, *Positive solutions of nonlinear singular two-point boundary value problems for second-order impulsive differential equations*, Applied Mathematics and Computation 196 (2008) 550–562.

[127] G. Infante, F. Minhós, P. Pietramala, *Non-negative solutions of systems of ODEs with coupled boundary conditions,* Commun. Nonlinear Sci. Numer. Simulation **17** (2012) 4952–4960.

[128] G. Infante, P. Pietramala, *Eigenvalues and non-negative solutions of a system with nonlocal BCs,* Nonlinear Stud. 16 (2009) 187–196.

[129] G. Infante, P. Pietramala, *Nonnegative solutions for a system of impulsive BVPs with nonlinear nonlocal BCs,* Nonlinear Analysis: Modelling and Control, Vol. **19**, No. 3, (2014) 413–431.

[130] G. Infante, P. Pietramala, *Existence and multiplicity of non-negative solutions for systems of perturbed Hammerstein integral equations,* Nonlinear Anal. Theor. Meth. Appl. 71 (2009) 1301–1310.

[131] G. Infante, F. Minhós, *Nontrivial solutions of systens of Hammerstein integral equations with firsrt derivative dependence,* Mediterr. J. Math. (2017) 14: 242. https://doi.org/10.1007/s00009-017-1044-1.

[132] B. Ilhan, I. Ozdemir, *Existence and asymptotic behavior of solutions for some nonlinear integral equations on an unbounded interval,* Electronic Journal of Differential Equations, 2016(271) (2016) 1–15.

[133] E. Izhikevich, *Dynamical Systems in Neuroscience: The Geometry of Excitability and Bursting,* The MIT Press, Computational Neuroscience, Cambridge (2007).

[134] T. Jang, *A new semi-analytical approach to large deflections of Bernoulli–Euler-v.Karman beams on a linear elastic foundation: Nonlinear analysis of infinite beams,* Int. J. Mech. Sci. 66 (2013) 22.

[135] T. Jang, H. Sung, S. Han, S. Kwon, *Inverse determination of the loading source of the infinite beam on elastic foundation,* Journal of Mechanical Science and Technology 22 (2008) 2350–2356.

[136] T. Jankowski *Nonnegative solutions to nonlocal boundary value problems for systems of second-order differential equations dependent on the first-order derivatives,* Nonlinear Anal. 87 (2013) 83–101.

[137] T. Jankowski, *First-order impulsive ordinary differential equations with advanced arguments,* J. Math. Anal. Appl. 331 (2007) 1–12.

[138] R. Jones, S. Stokes, B. Lockaby, J. Stanturf, *Vegetation responses to helicopter and ground based logging in blackwater floodplain forests,* Forest Ecology and Management 139 (2000) 215–225.

[139] J. Jost, *Postmodern Analysis,* 3rd edn., Springe-Verlag, Berlin (2005).

[140] T. Kajiwara, *A heteroclinic solution to a variational problem corresponding to FitzHugh-Nagumo type reaction-diffusion system with heterogeneity,* Commun. Pure Appl. Anal. 16(6) (2017) 2133–2156.

[141] P. Kang, Z. Wei, *Existence of positive solutions for systems of bending elastic beam equations,* Electron. J. Differential Equations **19** (2012).

[142] P. Kang, Z. Wei, *Three positive solutions of singular nonlocal boundary value problems for systems of nonlinear second-order ordinary differential equations,* Nonlinear Anal. Theor. Meth. Appl. 70 (2009) 444–451.

[143] Ö. Karabacak, P. Ashwin, *Heteroclinic Ratchets in Networks of Coupled Oscillators,* J. Nonlinear Science, 20 (2010) 105–129.

[144] V. J. Katz *A History of Mathematics. An Introduction*, 3rd edn., Addison-Wesley, New York (1998).

[145] E. Kaufmann, N. Kosmatov, Y. Raffoul, *A second-order boundary value problem with impulsive effects on an unbounded domain*, Nonlinear Anal. 69 (2008) 2924–2929.

[146] P. Kevrekidis, D. Frantzeskakis, *Solitons in coupled nonlinear Schrödinger models: A survey of recent developments*, Reviews in Physics 1 (2016) 140–153.

[147] A. Kneser, *Untersuchung und asymptotische Darstellung der Integrale gewisser Differentialgleichungen bei grossen Werthen des Arguments*, J. Reine Angen. Math 1, 116 (1896), 178–212.

[148] N. Komarova, A. Newell, *Nonlinear dynamics of sand banks and sand waves*, J. Fluid Mech., 415 (2000) 285–321.

[149] S. Konar, A, Biswas, *Properties of optical spatial solitons in photorefractive crystals with special emphasis to two-photon photorefractive nonlinearity*, Optical Materials 35 (2013) 2581–2603.

[150] L. Kong, *Homoclinic solutions for a higher order difference equation with p-Laplacian*, Indagationes Mathematicae 27 (2016) 124–146.

[151] W. Koon, M. Lo, J. Marsden, S. Ross, *Heteroclinic orbits and chaotic dynamics in planar fluid flows*, Chaos 10(2) (2000) 427–469.

[152] M. Krasnosel'skii, *Translation Operator Along the Trajectories of Differential Equations*, Nauka, Moscow, 1966 (Russian); English transl., Amer. Math. Soc., Providence, RI, 1968. MR 34:3012; MR 36:6688.

[153] E. Kreyszig *Introduction Funcional Analisys With Applications*, Jonh Wiley & Sons, New York, 1978.

[154] L. Kurz, P. Nowosad, B. R. Saltzberg, *On the solution of a quadratic integral equation arising in signal design*, J. Franklin Inst. B 281 (1966), 437–454.

[155] E.K. Lee, Y.H. Lee, *Multiple positive solutions of a singular Emden–Fowler type problem for second-order impulsive differential systems*, Bound. Value Probl., 2011, Art. ID 212980, 22 pp., 2011.

[156] E. Lee, Y.-H. Lee, *Multiple positive solutions of a singular Gelfand type problem for second-order impulsive differential systems*, Math. Comput. Model. 40 (2004) 307–328.

[157] E. Lee, Y.-H. Lee, *Multiple positive solutions of singular two point boundary value problems for second order impulsive differential equations*, Applied Mathematics and Computation 158 (2004) 745–759.

[158] Y.-H. Lee, X. Liu, *Study of singular boundary value problems for second order impulsive differential equations*, J. Math. Anal. Appl. 331 (2007) 159–176.

[159] A. Leung, *A semilinear reaction-diffusion prey-predator system with nonlinear coupled boundary conditions: Equilibrium and stability*, Indiana Univ. Math. J., 31 (1982) 223–241.

[160] H. Lian, W. Ge, *Solvability for second-order three-point boundary value problems on a half-line*, Applied Mathematics Letters, vol. 19 (2006) 1000–1006.

[161] H. Lian, W. Ge, *Existence of positive solutions for Sturm-Liouville bound-ary value problems on the half-line*, J. Math. Anal. Appl. 321 (2006) 781–792.

[162] H. Lian, H. Pang, W. Ge, *Triple positive solutions for boundary value prob-lems on infinite intervals*, Nonlinear Analysis 67 (2007) 2199–2207.

[163] H. Lian, H. Pang, W. Ge, *Solvability for second-order three-point boundary value problems at resonance on a half-line*, J. Math. Anal. Appl. 337 (2008) 1171–1181.

[164] H. Lian, P. Wang, W. Ge, *Unbounded upper and lower solutions method for Sturm-Liouville boundary value problem on infinite intervals*, Nonlinear Anal. 70 (2009) 2627–2633.

[165] X. Li, M. Bohner, C-K. Wang *Impulsive differential equations: Periodic solutions and applications*, Automatica 52 (2015) 173–178.

[166] X. Li, S. Song, *Impulsive control for existence, uniqueness and global sta-bility of periodic solutions of recurrent neural networks with discrete and continuously distributed delays*, IEEE Trans. Neural Netw. Learning Syst. 24(6) (2013) 868–877, doi: 10.1109/TNNLS.2012.2236352.

[167] X. Li, J. Wu, *Stability of nonlinear differential systems with state-dependent delayed impulses*, Automatica 64 (2016) 63–69.

[168] X. Li, S. Song, *Stabilization of delay systems: Delay dependent impul-sive control*, IEEE Trans. Automatic Control 62(1) (2017) 406–411, doi: 10.1109/TAC.2016.2530041.

[169] Y. Li, Y. Guo, G. Li, *Existence of positive solutions for systems of non-linear third-order differential equations*, Commun. Nonlinear Sci. Numer. Simulation 14 (2009) 3792–3797.

[170] G. Li-Jun, S. Jian-Ping, Z. Ya-Hong. *Existence of positive solutions for nonlinear third-order three-point boundary value problems*. Nonlinear Anal. 68 (2008) 3151–3158.

[171] L. Liu, L. Hu, Y. Wu, *Positive solutions of two-point boundary value prob-lems for systems of nonlinear second-order singular and impulsive differen-tial equations*, Nonlinear Analysis 69 (2008) 3774–3789.

[172] L. Liu, X. Hao, Y. Wu, *Unbounded Solutions of Second-Order Multipoint Boundary Value Problem on the Half- Line*, Boundary Value Problems, 2010 (2010), 15.

[173] Y. Liu, *Existence of Solutions of Boundary Value Problems for Coupled Singular Differential Equations on Whole Line with Impulses*, Mediterr. J. Math. 12 (2015), 697–716.

[174] X. Liu, H. Chen, Y. Lü, *Explicit solutions of the generalized KdV equations with higher order nonlinearity*, Appl. Math. Comput. 171 (2005) 315–319.

[175] Y. Liu, S. Chen *Existence of bounded solutions of integral boundary value problems for singular differential equations on whole lines*, Int. J. Math. 25(8) (2014) 1450078 (28 pages).

[176] L. Liu, P. Kanga, Y. Wub, B. Wiwatanapataphee, *Positive solutions of singular boundary value problems for systems of nonlinear fourth order dif-ferential equations*, Nonlinear Anal. 68 (2008) 485–498.

[177] Y. Liu, *Existence results on positive periodic solutions for impulsive functional differential equations*, Glasnik Matematicki 46(66) (2011) 149–165.

[178] Z. Liu, S. Guo, Z. Zhang, *Homoclinic orbits for the second-order Hamiltonian systems*, Nonlinear Anal. Real World Appl. 36 (2017) 116–138.

[179] E. Lombardi, *Oscillatory Integrals and Phenomena Beyond all Algebraic Orders with Applications to Homoclinic Orbits in Reversible Systems*, Springer-Verlag, Berlin, 2000.

[180] L. López-Somoza, F. Minhós, *Existence and multiplicity results for some generalized Hammerstein equations with a parameter*, https://arxiv.org/abs/1811.06118v1 (2018).

[181] H. Lü, H. Yu, Y. Liu, *Positive solutions for singular boundary value problems of a coupled system of differential equations*, J. Math. Anal. Appl. 302 (2005) 14–29.

[182] J. Maaita, E. Meletlidou, *Analytical homoclinic solution of a two-dimensional nonlinear system of differential equations*, J. Nonlinear Dynam. 2013 (2013), Article ID 879040, 4 pp., http://dx.doi.org/10.1155/2013/879040.

[183] N. Malhotra, N. Sri Namachchivaya, *Global bifurcations in externally excited two-degree-of-freedom nonlinear systems*, Nonlinear Dynam. 8(1) (1995) 85–109.

[184] C. Marcelli , F. Papalini, *Heteroclinic connections for fully non-linear non-autonomous second-order differential equations*, J. Differential Equations 241 (2007) 160–183.

[185] M. Meehan, D. O'Regan, *Multiple nonnegative solutions of nonlinear integral equations on compact and semi- infinite intervals*, Appl. Anal. 74 (2000) 413–427.

[186] F. A. Mehmeti, S. Nicaise, *Nonlinear interaction problems*, Nonlinear Anal. 20(1) (1993) 27-61.

[187] J. Melan, *Theory of Arches and Suspension Bridges*, Myron Clark Publ. Comp., 1913.

[188] Y. Mikhlin, T. Bunakova, G. Rudneva, N. Perepelkin, *Transient in 2-DOF Nonlinear Systems*, ENOC 2008, Saint Petersburg, Russia, June, 30-July, 4 2008.

[189] N. Milev, D. Bainov, *Stability of linear impulse differential equations*, Computers Math. Appl. 20 (12) (1990) 35–41.

[190] V. Milman, D. Myshkis, *On the stability of motion in the presence of impulses*, Siberian Math. J. 1(2) (1960) 233–237 (in Russian).

[191] F. Minhós, *Sufficient conditions for the existence of heteroclinic solutions for φ-Laplacian differential equations*, Complex Variables Elliptic Equ. 62(1) (2017) 123–134.

[192] F. Minhós, H. Carrasco, *Higher Order Boundary Value Problems on Unbounded Domains: Types of Solutions, Functional Problems and Applications*, World Scientific Publishing, New Jersey (2017).

[193] F. Minhós, *On the heteroclinic solutions for BVPs involving ϕ-Laplacian operators without asymptotic or growth assumptions*, Math. Nachr. (2018) 1–9, DOI: 10.1002/mana.201700470.

[194] F. Minhós, *Impulsive problems on the half-line with infinite impulse moments*, Lithuanian Math. J (2017) 57: 69, doi:10.1007/s10986-017-9344-5.

[195] F. Minhós, R. Carapinha, *Half-linear impulsives problems for classical singular φ-Laplacian with generalized impulsive conditions*, J. Fixed Point Theory Appl. 20(3) (2018) 117, DOI:10.1007/s11784-018-0598-2.

[196] F. Minhós, H. Carrasco, *Existence of homoclinic solutions for nonlinear second-order problems*, Mediterr. J. Math. 13 (2016) 3849–3861, doi:10.1007/s00009–016–0718–4.

[197] F. Minhós, R. Sousa, *Solvability of second order coupled systems on the half-line*, Lithuanian Math. J. (2019), 1–9. https://doi.org/10.1007/s10986-019-09419-y.

[198] F. Minhós, R. de Sousa, *Existence of homoclinic solutions for nonlinear second-order coupled systems*, J. Differential Equations 266 (2018) 1414–1428, doi.org/10.1016/j.jde.2018.07.072.

[199] F. Minhós, H. Carrasco, *Solvability of higher-order BVPS in the half-line with unbounded nonlinearites*, Dynamical Systems, Differential Equations and Applications, AIMS Proceedings, (2015) 841–850, doi:10.3934/proc.2015.0841.

[200] F. Minhós, R. de Sousa, *On the solvability of third-order three points systems of differential equations with dependence on the first derivatives*, Bull. Brazilian Math. Soc. (NS), DOI: 10.1007/s00574-016-0025-5.

[201] F. Minhós, *Solvability of generalized Hammerstein integral equations on unbounded domains, with sign-changing kernels*, Appl. Math. Lett. 65 (2017) 113–117.

[202] F. Minhós, I. Coxe, *System of coupled clamped beam equations: existence and localization results*, Nonlinear Analysis: Real World Applications 35 (2017) 45–60.

[203] F. Minhós, J. Fialho, *On the solvability of some fourth-order equations with functional boundary conditions*, Discrete Contin. Dyn. Syst. 2009(suppl) (2009) 564–573.

[204] F. Minhós, R. Carapinha, *On higher order nonlinear impulsive boundary value problems*, Dynamical Systems, Differential Equations and Applications, AIMS Proceedings, (2015) 851–860.

[205] F. Minhós, R. de Sousa, *Existence of solution for functional coupled systems with full nonlinear terms and applications to a coupled mass-spring model*, Differential Equations & Appl. 9(4) (2017) 433–452, doi:10.7153/dea-2017-09-30.

[206] F. Minhós, R. de Sousa, *Existence result for impulsive coupled systems on the half-line*, Revista de la Real Academia de Ciencias Exactas, Físicas y Naturales. Serie A. Matemáticas (2018), pp 1–14, https://doi.org/10.1007/s13398-018-0526-8.

[207] F. Minhós, R. de Sousa, *Localization results for impulsive second order coupled systems on the half-line and application to logging timber by helicopter*, Acta Appl. Math. (2018) 1–19, DOI: 10.1007/s10440-018-0187-9.

[208] F. Minhós, R. de Sousa, *Impulsive coupled systems with generalized jump conditions*, Nonlinear Anal. Model. Control 23(1) (2018) 103–119.

[209] F. Minhós, R. de Sousa, *Solvability of Coupled Systems of Generalized Hammerstein-Type Integral Equations in the Real Line*, Mathematics 2020, 8(1), 111; https://doi.org/10.3390/math8010111.

[210] F. Minhós, R. de Sousa, *Heteroclinic and homoclinic solutions for nonlinear second-order coupled systems with phi-Laplacians*, http://arxiv.org/abs/2003.14095.

[211] T. Moussaqui, R. Precup, *Multiple nonnegative solutions of nonlinear integral equations on compact and semi- infinite intervals*, Appl.Anal. 74 (2000) 413–427.

[212] R. Nagle, E. Saff, A. Snider, *Fundamentals of Differential Equations*, 8th edn., Pearson Education Limited, 2014.

[213] Z. Oniszczuk, *Damped vibration analysis of an elastically connected complex double-string system*, J. Sound Vibration 264 (2003) 253-271.

[214] D. O'Regan, B. Yan, R. Agarwal, *Solutions in weighted spaces of singular boundary value problems on the half- line*, J. Comput. Appl. Math. 205 (2007) 751–763.

[215] D. O'Regan, M. Meehan, *Existence Theory for Nonlinear Integral and Integrodifferential Equations*, Kluwer Academic Publishers, Dordrecht, 1998.

[216] P. Palamides, G. Galanis, *Positive, unbounded and monotone solutions of the singular second Painlev equation on the half-line*, Nonlinear Anal. 57 (2004) 401–419.

[217] H. Pang, M. Lu, C. Cai, *The method of upper and lower solutions to impulsive differential equations with integral boundary conditions*, Adv. Difference Equations 2014, 2014:183.

[218] J. Pava, F. Linares, *Periodic Pulses of Coupled Nonlinear Schrödinger Equations in Optics*, Indiana Univ. Math. J., 56(2), n. 847, n. 877, 2007.

[219] L. Perko, *Differential Equations and Dynamical Systems*, 3rd edn., Springer-Verlag, New York, 2006.

[220] W. Pogorzelski, *Integral Equations and their Applications*, Vol. I, Pergamon Press, 1966.

[221] C. W. Groetsch, *Integral equations of the first kind, inverse problems and regularization: A crash course*, J. Phys. Conf. Ser. 73 (2007) 012001, doi:10.1088/1742-6596/73/1/012001.

[222] I. Rachůnková, J. Tomeček, *State-Dependent Impulses: Boundary Value Problems on Compact Interval*, Atlantis Press, Paris, 2015.

[223] E. Radi, G. Bianchi, L. di Ruvo, *Analytical bounds for the electromechanical buckling of a compressed nanocantilever*, Appl. Math. Model. 59 (2018) 571–582.

[224] E. Radi, G. Bianchi, L. di Ruvo, *Upper and lower bounds for the pull-in parameters of a micro- or nanocantilever on a flexible support*, Int. J. Non-Linear Mech. 92 (2017), 176–186.

[225] M. Rahman, *Integral Equations and Their Applications*, WIT Press, Southampton, 2007.

[226] Randélović, L. Stefanović, B. Danković, *Numerical Solution of impulsive differential equations*, Facta Universitatis, Ser. Math. Inform. 15 (2000) 101–111.

[227] J. Rapp, T. Theodore, D. Robinson, *Soil, groundwater, and floristics of southeastern United States blackwater swamp 8 years after clearcutting with helicopter and skidder extraction of the timber*, Forest Ecology and Management 149 (2001) 241–252.

[228] H. Riecke, *Self-trapping of traveling-wave pulses in binary mixture convection*, Phys. Rev. Lett. 68 (1992) 301–304.

[229] V. Rodrigues, *Estudo das vibrações transversais em um sistema viscoelastico acoplado de duas cordas*, Santa Maria, RS, Brasil 2013.

[230] V. Rottschäfer, J. Tzou, M. Ward, *Transition to blow-up in a reaction–diffusion model with localized spike solutions*, Euro. J. Appl. Math. 28 (2017) 1015–1055. Cambridge University Press 2017, doi:10.1017/S0956792517000043.

[231] A. Samoilenko, N. Perestyuk, *Impulsive Differential Equations*, World Scientific Series on Nonlinear Science, Series A: Monographs and Treatises, World Scientific, Singapore, 14 (1995).

[232] S. Sauter, C. Schwab, *Boundary Element Methods*, Berlin: Springer-Verlag, 2011.

[233] W. Schiesser, *Method of Lines PDE Analysis in Biomedical Science and Engineering*, John Wiley & Sons, Inc., Hoboken, NJ, 2016.

[234] K. Shah, H. Khalil, R. Khan, *Investigation of positive solution to a coupled system of impulsive boundary value problems for nonlinear fractional order differential equations*, Chaos, Solitons Fractals 77 (2015) 240–246.

[235] A. Shahsavaran, *Lagrange functions method for solving nonlinear Hammerstein Fredholm-Volterra integral equations*, Appl. Math. Sci. 5(49) (2011) 2443–2450.

[236] L. Shen, R. Gorder, *Predator-prey-subsidy population dynamics on stepping-stone domains*, J. Theoret.l Biol. 420 (2017) 241–258.

[237] J. Shen, W. Wang, *Impulsive boundary value problems with nonlinear boundary conditions*, Nonlinear Anal. 69 (2008) 4055–4062.

[238] N. A. Sidorov, *Solving the Hammerstein integral equation in the irregular case by successive approximations*, Siberian Math. J. 51(2) (2010) 325-329

[239] N. A. Sidorov, *Explicit and implicit parametrizations in the construction of branching solutions by iterative methods*, Sbornik: Math. 186(2) (1995), 297, http://dx.doi.org/10.1070/SM1995v186n02ABEH000017.

[240] R. de Sousa, F. Minhós, *Coupled systems of Hammerstein-type integral equations with sign-changing kernels*, Nonlinear Anal. Real World Applications Volume 50, December 2019, Pages 469-483.

[241] R. de Sousa, F. Minhós, *On coupled systems of Hammerstein integral equations*, Boundary Value Problems, (2019) 2019: 7. https://doi.org/10.1186/s13661-019-1122-3.

[242] I. Stakgold, M. Holst, *Green's functions and Boundary Value Problems*. John Wiley & Sons, 3rd edn., 2011.

[243] I. Stamova, G. Stamov, *Applied Impulsive Mathematical Models*, Canadian Mathematical Society, Springer International Publishing Switzerland, 2016.

[244] C. Subrata, *Application and verification of deepwater Green function for water waves*, J. Ship Res. 45(3) (2001) 187–196.

[245] J. Sun, T. Wu, *Two homoclinic solutions for a nonperiodic fourth order differential equation with a perturbation*, J. Math. Anal. Appl. 413 (2014) 622–632.

[246] J. Sun, T. Wu, *Multiplicity and concentration of homoclinic solutions for some second order Hamiltonian systems*, Nonlinear Anal. 114 (2015) 105–115.

[247] I. Talib, N.A. Asif, C. Tunc, *Coupled lower and upper solution approach for the existence of solutions of nonlinear coupled system with nonlinear coupled boundary conditions*, Proyecciones J. Math. 35(1) (2016) 99–117.

[248] Y. Tian, W. Ge, *Positive solutions for multi-point boundary value problem on the half-line*, Journal of Mathematical Analysis and Applications, vol. 325, (2007) 1339–1349.

[249] S. P. Timoshenko, *Theory of elastic stability*, McGraw-Hill, New York, 1961.

[250] P. Torres, *Mathematical Models with Singularities*, Atlantis Press, Paris, 2015.

[251] E.O. Tuck, L.W. Schwartz, *A numerical and asymptotic study of some third-order ordinary differential equations relevant to draining and coating flows*, SIAM Rev. 32 (3) (1990) 453–469.

[252] Y. Wang, L. Liu, Y. Wu, *Positive solutions of singular boundary value problems on the half-line*, Appl. Math. Comput. 197, (2008) 789–796.

[253] W-X. Wang, L. Zhang, Z. Liang, *Initial value problems for nonlinear impulsive integro-differential equations in Banach space*, J. Math. Anal. Appl. 320 (2006) 510–527.

[254] A. Wang, J. Sun, A. Zettl, *The classification of self-adjoint boundary conditions: separated, coupled*, J. Funct. Anal. 255 (2008) 1554–1573.

[255] X. Wang, Y. Wang, *Novel dynamics of a predator-prey system with harvesting of the predator guided by its population*, Appl. Math. Model. 42 (2017) 636–654.

[256] J. Wang, S. Grushecky, J. McNeel, *Production analysis of helicopter logging in West Virginia: A preliminary case study*, Forest Products J. 55 (2005) 71–76.

[257] D. Wilczak, *Symmetric homoclinic solutions to the periodic orbits in the Michelson system*, Topological Methods in Nonlinear Analysis Journal of the Juliusz Schauder Center 28 (2006) 155–170.

[258] Workers' Compensation Board, *Safe work practices for helicopter operations in forest industry*, 2015 Edition (http://www.wcb.ny.gov/).

[259] J. Xu, Z. Yang, *Positive solutions for a system of nonlinear Hammerstein integral equations and applications*, J. Integral Equations Appl. 24(1) (2012) 131–147.

[260] B. Yan, D. O'Regan, R. Agarwal, *Unbounded solutions for singular boundary value problems on the semi-infinite interval: Upper and lower solutions and multiplicity*, J. Comput. Appl. Math. 197 (2006) 365–386.

[261] B. Yan, Y. Liu, *Unbounded solutions of the singular boundary value problems for second order differential equations on the half-line*, Appl. Math. and Comput. 147 (2004) 629–644.

[262] J. Yang, Z. Wei, *On existence of positive solutions of Sturm-Liouville boundary value problems for a nonlinear singular differential system*, J. Appl. Math. Comput., 217 (2011) 6097–6104.

[263] T. Yang, *Impulsive Control Theory*, Springer-Verlag Berlin, 2001.

[264] Y. Wang, *Some fourth order equations modeling suspension bridges*, Politecnico di Milano, Ph.D. dissertation, 2015.

[265] Z. Yang, Z. Zhang, *Positive solutions for a system of nonlinear singular Hammerstein integral equations via nonnegative matrices and applications*, Positivity 16(1) (2012) 783–800, DOI:10.1007/s11117-011-0146-4.

[266] J. Yeh, *Real Analysis: Theory of Measure and Integrations*, 2nd edn., World Scientific Publishing, Singapore (2006).

[267] F. Yoruk, N. Aykut Hamal, *Second-order boundary value problems with integral boundary conditions on the real line*, Electronic J. Differential Equations 2014(19) (2014) 1–13.

[268] M. Zarebnia, S. Khani, *Numerical solution of Hammerstein integral equations by using quasi-interpolation*, Eng. Technol. Int. J. Math. Comput. Sci. 7(4) (2013).

[269] E. Zeidler, *Nonlinear Functional Analysis and Its Applications: Fixed-Point Theorems*, Springer, New York (1986).

[270] V. Zelati, P. Rabinowitz, *Heteroclinic solutions between stationary points at different energy levels*, Topol. Methods Nonlinear Anal. J. Juliusz Schauder Center Volume 17, 2001, 1–21.

[271] T. Zhang, Y. Jin, *Traveling waves for a reaction-diffusion-advection predator-prey model*, Nonlinear Anal. Real World Applications 36 (2017) 203–232.

[272] X. Zhang, L. Liu, Y. Wu, *Existence and uniqueness of iterative positive solutions for singular Hammerstein integral equations*, J. Nonlinear Sci. Appl., 10 (2017) 3364–3380.

[273] X. Zhang, L. Liu, Y. Wu, *Global solutions of nonlinear second-order impulsive integro-differential equations of mixed type in Banach spaces*, Nonlinear Anal. 67 (2007) 2335–2349.

[274] F. Zhu, L. Liu, Y. Wu, *Positive solutions for systems of a nonlinear fourth-order singular semipositone boundary value problems*, Appl. Math. Comput. **216** (2010), 448–457.

[275] D. Zill, M. Cullen, *Differential equations with Boundary-Value Problems*, 7th edn., Brooks Cole, 2008.

[276] F. Zun-Tao, L. Shi-Da, L. Shi-Kuo, L. Fu-Ming, X. Guo-Jun, *Homoclinic (Heteroclinic) Orbit of Complex Dynamical System and Spiral Structure*, Commun. Theor. Phys. (Beijing, China) 43(4), (2005) pp. 601–603.

Index